ADDITIONAL PRAISE FOR *FIRESTARTERS*

"As a businesswoman who has been involved with multiple ventures, I strongly appreciate a book with solid advice, new ways of thinking, and inspiring examples."

—Anne Zimmer, CFO of Black Button Distilling

"Oftentimes Innovators get stuck in the conceptual stage and never get to market. *Firestarters* provides meaningful strategies on how to transition from thought to action."

—Fenorris Pearson, former VP of global consumer innovation for Dell and author of *How to Play the Game at the Top*

"Hits home big-time! A must-read! I have always proudly worn the badge of Instigator, and *Firestarters* validates what I have done and what I plan on doing in the future."

—Carol Fitzgerald, executive vice president of Life Medical

"My world is filled with entrepreneurs embracing opportunities. *Firestarters* provides useful advice and examples about how to eliminate Extinguishers that prevent people from achieving their dreams."

—Amy Castronova, president of Novatek Communications, Inc., and trainer for Entrepreneurs' Organization

"As soon as I read the title and overview of *Firestarters*, I immediately thought, 'They're talking about folks like me!' An athlete from a blue-collar family with aspirations to play professional football, after graduating with a degree in print journalism in 1991, I started a small publishing house that has now ballooned into twenty-eight books and counting, including six *New York Times* bestsellers, all starting out of a studio apartment filled with hundreds of books and little furniture. Now my question becomes, 'How do you keep the fire lit or reignite that fire when or if it ever gets snuffed out?' The answers are all here in *Firestarters*, a must-read for the new millennium."

—Omar Tyree, *New York Times*–bestselling author and visionary of *The Equation: The Four Undisputable Components of Success*

"With *Firestarters*, the authors take the meaning of entrepreneurship to a whole new level."

—Kathie Okun, president of Okun Financial Group, Inc.

FIRESTARTERS

FIRESTARTERS

HOW INNOVATORS, INSTIGATORS, AND INITIATORS CAN INSPIRE YOU TO

IGNITE YOUR OWN LIFE

RAOUL DAVIS JR.
KATHY PALOKOFF
PAUL EDER

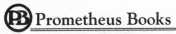
Prometheus Books

59 John Glenn Drive
Amherst, New York 14228

Published 2018 by Prometheus Books

Cover image © Weerachai Khamfu/Shutterstock
Cover design by Liz Mills
Cover design © Prometheus Books

Trademarked names appear throughout this book. Prometheus Books recognizes all registered trademarks, trademarks, and service marks mentioned in the text.

Inquiries should be addressed to
Prometheus Books
59 John Glenn Drive
Amherst, New York 14228
VOICE: 716–691–0133 • FAX: 716–691–0137
WWW.PROMETHEUSBOOKS.COM

22 21 20 19 18   5 4 3 2 1

Library of Congress Cataloging-in-Publication Data

Names: Davis, Raoul, 1979- author. | Palokoff, Kathy, 1953- author. | Eder, Paul, 1980- author.
Title: Firestarters : how innovators, instigators and initiators can inspire you to ignite your own life / by Raoul Davis, Kathy Palokoff, and Paul Eder.
Description: Amherst, New York : Prometheus Books, 2018. | Includes index.
Identifiers: LCCN 2017034253 (print) | LCCN 2017043829 (ebook) | ISBN 9781633883482 (ebook) | ISBN 9781633883475 (pbk.)
Subjects: LCSH: Creative ability in business. | Entrepreneurship. | Technological innovations. | Success in business.
Classification: LCC HD53 (ebook) | LCC HD53 .D384 2018 (print) | DDC 650.1—dc23
LC record available at https://lccn.loc.gov/2017034253

Printed in the United States of America

CONTENTS

ACKNOWLEDGMENTS

PAUL'S ACKNOWLEDGMENTS

First and foremost, I want to thank my wife, Marci, and my sons, Chase, Brady, and Dawson, for being continuous sources of joy, support, and just the right amount of craziness. I look forward to discovering new traditions together as we help each other build new fires.

I also want to recognize my colleagues from the Center for Organizational Excellence and my professors from University of Delaware and Loyola University Maryland for inspiring my integration of psychological research with practical applications.

And to my parents, my extended family, and my friends who have supported me over the years, I am thankful for you all.

KATHY'S ACKNOWLEDGMENTS

When you have lived a rich and long life like I have, acknowledgments are a strange thing. It seems like you almost have to acknowledge the universe since the people who touched and molded you are sometimes known and other times invisible. I am now sympathetic to all those Oscar winners. So I will forgo names, but you absolutely know who you are.

I thank my family and friends who are my world and my well of support. You never let it go dry and fill it with laughter, music, and wisdom. You hold my heart and have shaped me into who I am today.

My parents, sisters, sons, daughters, and grandchildren inspire me to be a Firestarter. My teachers and friends—old and new—ignite my spirit, and my sweetheart fuels my fire. My colleagues, clients, and partners in multiple business and social ventures along with peer-to-peer groups like Women Presidents' Organization. Great Dames and Ballers accelerate my growth and kept the Extinguishers away. My very cool coauthors spark things in me I didn't

9

know existed. And I thank those of you who read and comment on this book and help the flames burn inside me and inside yourself.

Most importantly, thank you God for giving me life. To be alive is to be able to spend one more day living my passion.

RAOUL'S ACKNOWLEDGMENTS

I want to thank my father, Raoul Davis Sr., who's been a model to me of what a Firestarter is. He founded the Urban League of Long Island, New York, and he worked with gang members in New York City to put down their guns and work together to formulate a jazz band. To my sisters, Rianna-Davis Gaetano and Tierney Hogan, thanks for your support and inspirational example.

A special thanks to Shaun Lewis, founder of N Group Consulting Services who was instrumental in helping us land some of our most important interviews for the book. Shaun is an influential leader who is a Firestarter behind the scenes and who helps keep the business world spinning. To the Ascendant Group team, Kimberly Reed, Louis Lautman, Richelle Payne, David Novak, Davida Pitts, Annika Murray, Merilee Kern, Cheri Swalwell, Astrid Peters, Ezz El Nattar, and Sophia Nelson. Aleen Zakka and the many other partners like Karen Leland, Holley Richardson, and Charles Smith who helped build a world-class CEO branding firm, thank you. To my wife, Demetreecia Odom Davis, thank you for putting up with this Firestarter; I know it isn't always easy; however, we've always worked together to endure. To my talented nephew Langston, all things are possible. And finally, my son Austin Raoul Davis—you are already growing up to be a Firestarter; I hope this book helps you live to your full potential and you're able to live a life that paves others to reach their Firestarter potential.

To all the readers, thank you for buying this book. We hope it inspires you, causes you to reflect, and most importantly drives you to action. We would love to hear from you.

Section 1

UNDERSTANDING FIRESTARTERS

Chapter 1

WHY WE BECAME CURIOUS ABOUT FIRESTARTERS

Some people start jobs; others start companies. Some wear clothes; others start trends. Some wish for power; others become world-class leaders. Some join a cause; others create movements.

Some fan the flames; others spark the fire. We call these unique individuals *Firestarters*. They are the Innovators who create things, Instigators who disrupt things, and Initiators who start things. Together, they make the world a much better place.

They are individuals who spark change globally and who are powerful forces in their local communities. Some focus on being leaders in established disciplines. Others break ground in new areas of discovery. Some use social ventures and nonprofits to improve the lives of others. Others build strong companies that change the world.

So, who are these company-building, cause-creating, trend-starting, influence-wielding renegades? They are athletes, writers, mothers, fathers, scientists, philosophers, business executives, entrepreneurs, politicians, doctors, inventors, teachers, and people from a multitude of other walks of life.

All are Firestarters. Within ten minutes of talking with them, you can tell they are different than your average person. You discover that they inspire all of us to be just a little better than we are.

How do Firestarters make such an impact? What makes them so different from anyone else? What can we learn from them so that we can become Firestarters ourselves? What is the difference between those who dream and those who do?

We turned to multiple sources of wisdom for the answers. First, we examined the extensive research that is being conducted across disciplines in motivation and entrepreneurship. Second, we turned to the Firestarters themselves. We wanted to learn from the nuggets and themes that made up their unique stories and perspectives. Finally, we reviewed the lives of Firestarters throughout history.

Because we are so passionate about this topic, we talked to and read about hundreds of people who start ventures, ignite communities, and further causes. We searched for elements that make up the Firestarter experience—Igniters that start the fire, Fuels that feed the fire, and Accelerants that spread the fire. We also examined Extinguishers, the elements that threaten the fire.

WHY FIRESTARTERS?

Why do we want to know how Firestarters ignite their lives and yours? Because so many people have dreams, yet so few are willing to take action. We need more Firestarters willing to make things happen. We want to find the secret sauce, learn the recipe, and pass that knowledge on to create a blazing fire across the globe.

Without Firestarters, the world would still turn. However, economies would not thrive in quite the same way. Social reform would not happen as quickly. New frontiers would remain unexplored for longer periods of time. Over a billion people wouldn't be connected on Facebook, and there wouldn't be such a big deal over 140-character tweets. Your lifespan would likely be shorter by a few years, and it definitely would not be filled with as many modern wonders.

In a world filled with complex and daunting problems, it would be wonderful to have a large army of Firestarters attacking challenges such as disease, hunger, poverty, violence, wildlife preservation, water and air quality, energy shortages, terrorism, racism, crimes against women, and the host of issues facing our communities and nations.

What exactly does it take to become a Firestarter? In this book, we aim to peek into their brains, find the Firestarter switch, and teach you how to turn it on. The insights we share will enable you to ignite the movement inside your heart that is waiting for a spark.

Here's what we believe. There are characteristics that individuals possess that can be amplified in the right situations to help them on the road to becoming a Firestarter. For example, Firestarters exert freely chosen effort that is not undermined by persistent ridicule. Firestarters find their own drive and motivation from the effort expended, regardless of the outcome achieved. Accordingly, these individuals grow and become more resilient to potential setbacks.

This book is not about personality per se as that is a less malleable construct. We believe that being a Firestarter is accessible to all. Do Firestarters have a certain DNA? We believe so, but not in the genetic sense. The factors that distinguish Firestarters are much more situational than genetic. The DNA of a Firestarter is found in the unique way in which an individual interacts with his/her environment rather than the unique way genes are arranged.

Firestarters systematically seek out experience and learn from situations that reward them for high levels of productive effort. They perceive a high degree of control over their thoughts and actions and freely choose endeavors that allow them to capitalize on the type of effort that is most personally motivating.

When the right combination of personal attitudes meets the right situation, they become individuals whose passion and motivation for excellence dwarf those of the typical person. The drive for excellence for the Firestarter is a lifestyle, not an outcome. Consequently, they tend to outpace and outachieve others in the long run, even those who possess an outcome-focused mindset.

THREE EXAMPLES OF FIRESTARTERS

As stated, we have talked to and read about hundreds of Firestarters since beginning this project. One of the first questions people ask us about Firestarters is, "Do you mean 'entrepreneurs'?" The answer is *yes* and *no*. Entrepreneurs obviously make up a subset of Firestarters, but Firestarters are so much more. Here are three well-known names that epitomize what we call Firestarters and how we define Innovators, Instigators, and Initiators.

Innovator Elon Musk revolutionized online payments leading to the establishment of PayPal. If that was not enough, he then founded Tesla, which has modernized the car industry by successfully putting some of the first cars not dependent on gasoline on the road.

For most people, this level of innovation would be their life's work; however, Musk was just getting started. His next venture was his most ambitious yet. Space X was created to take humans to Mars. Asked why he created Space X, Musk decisively answered, "[B]ecause if we can become an interplanetary species, the likelihood of humanity ever being wiped out by an extinction level event decreases significantly."[1]

Mother Teresa is often known for her kindness, which is a stature she definitely earned. However, Mother Teresa was an Instigator and a missionary renegade. She was never shy to make her voice known to the powerful and elite in the world. She would call on them to recognize their guilt for the crime of poverty they created. She was slow to compromise and quick to fight for those without a voice.[2]

Mother Teresa built an organization with thousands of people running orphanages that cared for alcoholics, the blind, disabled, flood victims, and others considered the most vulnerable in society. After a person's death, there is a normal five-year standard before a nomination for sainthood is given. In Mother Teresa's case, this time frame was waived, and even her critics honored her for her accomplishments.

John Adams, the second president of the United States, was the ultimate Initiator. Adams wasn't known for his creativity but instead for his willingness to act and be first. He famously agreed to serve as lawyer for the British soldiers charged in the Boston Massacre out of fierce devotion to the principle of law.[3]

During the Continental Congresses, Adams was on more committees than anyone else—a staggering ninety, while chairing many of them.[4] Adams even became the first president to live in the White House. Adams's focus was on getting things done.

PULLING BACK THE CURTAIN

We'd like to take a few moments to tell you a bit about ourselves and how we came together on this project. It's a good example of Firestarters at work. For several decades, each of us individually observed amazing people in action. Our own careers in social psychology, executive and CEO branding, entrepreneurship, and marketing put us in contact with Firestarters from all walks of life.

Kathy Palokoff first fell in love with the term after founding several diverse businesses and social ventures, forging strong bonds with other highly successful women business owners, teaching in one of the world's leading advertising programs, mentoring hundreds of entrepreneurs, and being inspired by her own adventure-seeking sons and daughters. While searching for a way to better define who she was and what she did, one of her mentors, Executive Coach Extraordinaire Bobbie Goheen, described Kathy as "a real firestarter." That statement started her on her journey.

A few years later, she received a phone call from Betty Hines, her Baltimore chapter chair of the Women Presidents' Organization (WPO) stating that she was looking to bring in Raoul Davis, an external branding speaker for an upcoming meeting. Betty wanted to ensure this didn't conflict with Kathy's marketing business. After a high-energy dialogue, Kathy and Raoul decided to present a WPO session together, and Raoul became enthralled with Kathy's description of Firestarters.

Raoul remembered his own journey, when as a twenty-two-year-old vice president in student government, he asked a question to a speaker's bureau based in Los Angeles that changed his life. He asked if the company Speakers Etcetera had any internships. They said no, but he asked them to "give him one anyway." The answer eventually became yes. Raoul used an innovative sales model to become the top salesperson for the company by age twenty-four, despite only working fifteen hours per week while in college and graduate school.

Raoul went on to launch an innovative speaker management company, which was one of a kind at the time, before repositioning the company as a personal branding firm and eventually the most integrated CEO branding firm in the world. His company has held this positioning since 2009 and recently became a founding member of the Forbes Agency Council for PR & Advertising. Raoul has also gone on to participate in White House forums on entrepreneurship and has been part of the innovative team at Three Squared, one of America's leading companies revolutionizing the housing industry by building stronger walls utilizing shipping container materials.

After working together and developing the initial concept of Firestarters, Kathy and Raoul realized there was something missing. While their practical knowledge served them well, both agreed layering in a more academic and research-based approach would complete the model. Raoul turned to his roommate from graduate school, Dr. Paul Eder. Paul was working for the Center for Organizational Excellence in the Washington, DC, area. Paul earned his PhD in social psychology and has spent years integrating academic research findings on motivation and performance in practitioner settings.

As the three of us began to work together, we realized Kathy is an Initiator, Raoul is an Innovator, and Paul is an Instigator. Though all of us had experiences with each type in some situations, we definitely had defaults. Because our personal Firestarter experiences differed, our perspectives and ways of working together varied deeply. We developed a deep respect for each other and what each author brought to the table. The interviewees, the interpretations of the research, and the anecdotes discussed are a result of the three of us working, arguing, and laughing together as we integrated our life experiences and knowledge.

HOW TO READ THIS BOOK

We recommend reading *Firestarters* from top to bottom to understand concepts and the language we are using. However, you can also focus on individual sections.

If you are curious as to whether or not you have Firestarter potential and if you are an Innovator, Instigator, or Initiator, you might want to take a quick jump to chapters 35 and 36. Two exercises presented in those chapters will provide some insights about yourself.

In this book, we discuss different types of Firestarters as well as elements that influence them. We draw from historical and cultural anecdotes, as well as the multidisciplinary research to explore major theories and ideas about what makes Firestarters tick. We present unique and rich stories of inspiring individuals. In addition to giving you a look into their lives, we have designed their profiles to make it easier to understand their makeup as Firestarters.

Selecting the Firestarters for inclusion in this book was not easy, but it was intentional. We sought examples of different ages, public recognition, genders, cultural backgrounds, industries, impact, and other factors because we wanted to show that Firestarters are everywhere.

At the end of the book, we summarize our conclusions and recommendations. We have included advice on how to overcome threats and support the Firestarters in your life. We also explain why the world needs more Firestarters, how to make that happen, and the ripple effect they have through time.

Finally, we offer several activities and exercises to help you ignite your lives in new ways. You will find concrete advice to enhance yourself as a Firestarter, guide you on your journey to become a Firestarter, and help you support the Firestarters in your life.

And if you want even more, go to www.firestartersmedia.com.

SPARK YOUR THINKING

1. Do you believe everyone has Firestarter potential? Why?
2. What do you think makes someone a Firestarter?
3. Who do you consider three of the top Firestarters in history?
4. Who do you consider the top three Firestarters who have inspired you personally?
5. Can you see yourself having Firestarter potential? Why?

Chapter 2

THE FIRESTARTER FRAMEWORK

The Firestarter Framework focuses on themes and patterns that ran through our interviews and research. Through this framework, you will see Firestarters are not superheroes. They experience threats and fears. They feel human emotions and face setbacks like everyone else. But they push forward as issues arise.

You will understand how they seek out situations where their impact can be maximized. They find ways to change the situation when the available options aren't a match for their unique drive. This person-by-situation interaction intrigued us, and we sought a way to capture this concept fully to share it with you.

OVERVIEW OF THE FIRESTARTER FRAMEWORK

There are three primary types of Firestarters—Innovators, Instigators, and Initiators—who are driven by differing motives: creating, disrupting, or starting things. Firestarters set these motives into action by combining various elements from three buckets that we describe as Igniters (what lights the fire), Fuels (what feeds the fire), and Accelerants (what spreads the fire).

Igniters are situational motivators that drive you forward in a given moment. Fuels are environmental resources that provide energy for your journey. Accelerants are context-driven actions that propel you forward and feed off the momentum of your Igniters and Fuels.

These elements represent a complex interaction of person, situation, environment, and action. This perspective is consistent with our interviews as well as social cognitive theories supported by research in behavioral psychology. This line of research highlights the importance of personal factors, environment, and behavior working in tandem to influence outcomes.[1]

The Firestarter Framework

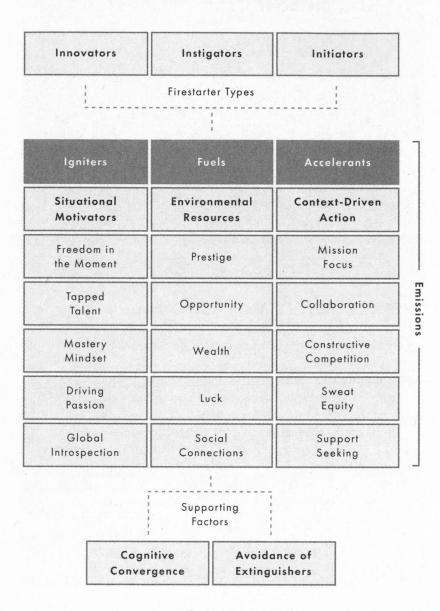

Fig. 2.1. The Firestarter Framework.

Firestarters apply unique combinations of the elements from the buckets to ignite their lives and make a difference in the world. The combinations chosen by any individual represent unique approaches to life, and some combination of elements from all three buckets is usually required for fires that burn the hottest and flourish the longest.

Along their journey, Firestarters apply the lessons from their past in a process we call *cognitive convergence*. They allow successes from their lives to converge in the moment. They draw on lessons, even from seemingly dissimilar events, to find the best way to apply them in a given situation. They download enough optimism to power through their obstacles.

Firestarters avoid or overcome elements of the environment and themselves that threaten to extinguish their flame. We refer to these elements as Extinguishers. People can be extinguished at the point of ignition, or they can run out of fuel after a fire has started. We all face these threats, and the Firestarters we interviewed shared powerful lessons on how they can be overcome or avoided.

Additionally, the impact of many Firestarters can be felt exponentially. Impact creates a ripple in time like a stone skipping across a lake. We refer to these ripples as Emissions that inspire others to act boldly and result in expansion of the Firestarter's potential legacy.

UNDERSTANDING FIRESTARTER "TYPES"

We debated precisely how to describe Firestarter "types"—Innovators, Instigators, and Initiators. They aren't "personality" types, although it is likely that one's personality would drive a particular Firestarter to seek out or to enjoy one of the three roles more than the others. Instead, the types represent approaches to the world and how we choose to grapple with the unknown. They are our default "tendencies."

All Firestarters don't have the same occupation or interests, but the three Firestarter types cross profession and field. Entrepreneurs are one subset of Firestarters. A fiery advocate for a cause can also be a Firestarter. A person with a disability fighting against all odds to succeed is a Firestarter. A middle manager fighting for a product or marketing campaign can be a Firestarter. Throughout their journeys, Firestarters often find ways to identify with all three types in themselves. By opening themselves in this way, they are more able to adapt in the face of challenges to accomplish what's needed.

In an *Academy of Management Review* paper, Melissa Cardon and her colleagues explained the concept of entrepreneurial passion using various *role identities*.[2] The three role identities described included the inventor, the founder, and the developer.

In the Firestarter Framework, the inventor role identity aligns with our concepts of Innovator and Instigator. The founder role identity aligns with the Initiator. Regarding the developer, we view this role identity to be intertwined with our concept of Accelerants, which we will describe later in more detail.

Consistent with our interviews, Cardon and her fellow theorists proposed that multiple role identities could be adopted equally by an entrepreneur. Alternatively, individuals may find one identity to be more salient or meaningful than others.

Sheldon Stryker and Peter Burke are researchers in the field of identity theory who have described identity from a hierarchical perspective.[3] Individuals can hold many simultaneous identities (i.e., I am a woman. I am a water polo player. I am a world-class Scrabble competitor). However, the salience of each can change over time and across situations. Simply put, role identities represent distinct but organically flowing motivations for igniting yourself and others.

The role identity perspective appeals to us. It addresses the concerns we had with the Firestarter types having the ability to be both stable and transient. Individuals may have a dominant type, but in the moment, Firestarters are flexible. They recognize situational needs and are able to flow into the accessible role identity most relevant to overcome emergent challenges. They are able to put on their "Innovator hat" in the morning for creative problem-solving and then transition to an Initiator at night when work just needs to get done.

Matthew Toren, founder of YoungEntrepreneur.com, also gives us perspectives on Firestarter types in an article he wrote about why people become entrepreneurs.[4] Three of the reasons included the following: having unconventional ideas, wanting to do things, and wanting to change the world. Sound familiar? We believe these three reasons go beyond entrepreneurs and correspond to Innovators, Instigators, and Initiators.

INNOVATORS CREATE THINGS

Innovation is enormously important. It's the only insurance against irrelevance.

—Gary Hamel, most reprinted author in *Harvard Business Review* history[5]

We say that Innovators create things. But this description has a lot of latitude. What does it mean to create? Does it always have to be a pure creation? Is a simple line drawn on a piece of paper an innovation? What if the line connects to other lines to make a picture? Does the picture have to be different in some way from other pictures?

Researchers at the very least agree that innovation involves two stages: developing ideas and implementing them.[6] Tom Freston, former president of Viacom, defined innovation as "taking two things that already exist and putting them together in a new way."[7] This is partially true. The *creativity* involved in innovation can result in an entirely new idea or the effective hijacking of old ideas for new purposes.

In the mid-1800s Charles Babbage invented a mathematical counting machine. His protégé, Ada Lovelace, translated a description of his invention from French to English. In the process, she compiled a notes section that was longer than the written text itself.[8] Within the notes, she theorized on the use of such machines for carrying out what computer scientists would later call an algorithm. Though lesser known in her time, Ada's work came to light and was honored by computer scientists over a century later.

Ada Lovelace was an Innovator. She saw a potential in Babbage's machine that he hadn't even considered. Innovators recognize potential in all facets of life and act on this potential. Ada's translation task did not require an extensive notes section, but her intuition drove her further than the task itself.

Innovators find ways to tweak things we already use, buy, or write about to make them seem new. For example, John Sculley, one of our featured Firestarters, made his mark before becoming CEO of Apple with the Pepsi Challenge.[9] He created a new and memorable way for people to differentiate two products that are similar—Pepsi and Coke.

As innovation often involves the effective implementation of creative ideas, we explored the pertinent research on how creativity happens. Teresa Amabile's componential model of creativity is one of the most researched and supported frameworks for creativity. This model holds creativity to be a product of intrinsic interest, domain-relevant skills, creativity-relevant processes, and the social context.[10] Other researchers also propose that individual creativity (and organizational creativity at the macro level) results from a combination of the person, situation, and group context.[11] Why is this combination important?

Innovators don't exist in isolation. In order to innovate, something has to enter their consciousness as having a need to be created. Alexander Graham Bell didn't invent the telephone just because he thought it "would be cool"— though he likely did think that. Rather, there was a cultural zeitgeist and industrial revolution that had recently brought the telegraph to life. A communications renaissance was occurring, and Bell got caught in the wave.

People around him were interested in communicating across distances. People around him were complaining about difficulties with existing communications mechanisms. Scientists were discovering new qualities of sound and how it traveled. Was Bell a great inventor? Yes. Outside of his time period and context, he likely would have invented many other nifty things. But in the situation, his personal attributes and context-driven behavior interacted to produce his moment of ignition in the late 1800s.

Think of the rise of Nintendo in the 1980s. Their successful video game designer Shigeru Miyamoto envisioned a game (*Donkey Kong*) that featured what some describe as a love triangle between a carpenter, an ape, and a girl.[12] That carpenter evolved into a plumber who had to rescue a princess from a fire-breathing reptile in the Super Mario Bros. franchise.

Admit it. These aren't ordinary ideas. It's very likely that even if you had this vision, you would have a difficult time selling it to your employer. As an Innovator, Miyamoto found ways to differentiate his ideas in the marketplace, and his unconventional approach to gaming has had ripples that continue to impact gamers more than thirty years later.

Innovators are owners of the situation. They own it because they create it—quite literally. They embrace the world as it should match the vision in their heads. And when something is missing from that vision, they fill the gap.

In chapter 27, we tell the stories of many Innovators. You may want to read a few now to gain a flavor of how Innovators—known and unknown—can inspire you to ignite your own life.

INSTIGATORS DISRUPT THINGS

> *Cautious, careful people, always casting about to preserve their reputations . . . can never affect a reform.*
> —Susan B. Anthony

Firestarters are generally not a silent breed. This is true of Instigators more than any other type. They speak out; they spar; they mentally joust; they wreak havoc (just ask Raoul and Kathy about Paul's contributions to this book). And through it all they make an impact that lasts.

In the course of our interviews, the term "Firestarter" was challenged. One interviewee even proposed that "Firefighter" was more appropriate. We found this dichotomy intriguing. How exactly could igniting one's life feel like the opposite? Yes, we understand that the concept is more important than the terminology, but it's an interesting perspective to explore in order to ensure the Firestarter Framework resonates appropriately.

Thinking about the term firefighter from an Instigator perspective makes it a little clearer. As a change agent, a Firestarter with a salient Instigator identity seeks out inefficiencies, inaccuracies, and generally anything that doesn't seem right. Once found, they attempt to eradicate the "badness" through bold action. This can feel like a firefight.

When we discuss Firestarters, we're referring to the unification of the person and situation—how the person can boldly transform a situation to make an impact. Instigators do this. They start new fires and transform the nature of other fires that have raged out of control.

Change is hard. In order to be successful at changing something, you have to know a lot about the world as it is. Think about it. The most successful arguments for change always incorporate the perspective of those whose behavior, attitudes, or habits you want to change. Instigators have to be experts at both the *As-Is* and *To-Be* worlds.

When Christopher Columbus sought funding for his voyages, he wanted to prove the existence of a new trade route from Europe to the Far East. Such a proposition was viewed as preposterously risky by the majority of the royalty in Europe whom he approached.[13] For years, he honed his pitch. Finally, he arrived in the presence of the monarchs of a newly unified Spanish empire, Ferdinand and Isabella. While initially skeptical, they soon succeeded in securing their territory from the Spanish Moors. Without a war to drain resources, they had broader openness to thoughts of expanding their empire in other regions.

Columbus's pitch finally worked, and he even secured many lofty promises to increase his own wealth and titular prestige should his voyage be successful. Columbus was an Instigator. He looked to change the way Europe thought about trade (and enrich himself in the process, of course). In order to succeed, he had to understand the world he wanted to change. The pitch had to resonate with the right people at the right time.

A far cry from Christopher Columbus, think about Tupac Shakur. The embattled hip-hop artist and actor was considered by many to be one of the most prolific voices of social justice in the 1990s. He continuously battled personal demons that threatened to undercut his message. Tupac was 100 percent an Instigator, and one of his goals was to inspire others to become more self-aware and figure out the change they want to make.

The lyrics from one of Tupac's most famous songs, "Hail Mary," demonstrate the ability he had to ignite change while simultaneously having a duplicitous nature: in "Hail Mary" he points out how the promise of many youths is wasted in penitentiaries. In his lyrics, Tupac makes a plea for social change while admitting his faults and demonstrating the difficulty he and others have with doing the right thing.[14] Even decades after his death, his influence permeates the music industry.

Instigators change the world in small ways every day. They lash out at their surroundings because they know things can be done better. They shape the world one sharp critique at a time.

In chapter 28, a variety of Instigators give us a peek at what makes them the change agents of our world. Feel free to flip ahead and take a few moments to get inspired.

INITIATORS START THINGS

> It must be remembered that there is nothing more difficult to plan, more doubtful of success, nor more dangerous to manage than a new system. For the initiator has the enmity of all who would profit by the preservation of the old institution and merely lukewarm defenders in those who gain by the new ones.
> —Niccolò Machiavelli

When we investigated the concept of Initiators, we felt that this group often got the short end of the perception stick. Initiators are constantly in "go-go-go" mode. They see a problem, own it, and solve it. But their solutions often aren't sexy. Sexy is the realm of Innovators and Instigators.

Initiators move when others don't. For them, a worthy cause doesn't inspire them to donate to a charity; it triggers the need to start their own charity. They run government agencies. They volunteer for the PTA. They organize other soccer moms in creating an efficient carpool plan. They create checklists and checklists for their checklists to ensure i's are dotted and t's are crossed.

Initiators are extraordinary taskers across domains. They do what needs to be done practically. Idealism is put on hold in favor of the performance of actions that are needed in the moment.

Take Clara Barton for example. She always did what was needed. She tended to a sick brother. She opened a free public school at a young age.[15] When the American Civil War erupted, Clara was not content to sit on the sidelines. She took action as an independent nurse and earned a reputation as an "angel of the battlefield." After seeing the success of an aid organization during a European visit after the war, she returned to the United States and eventually founded the American Red Cross Society. Clara Barton was an Initiator. Her life was dedicated to helping where help was needed, and her help was often in a leadership role.

Business writer Paul Brown summed up the Initiator type nicely in a 2013 Forbes.com post: "They spotted a need, and they did something about it."[16] Starting something is akin to *doing*, which is an active way to approach the world. Initiators live by the mantra "nothing was ever accomplished by standing still."

In chapter 29, you will find insights from Initiators across disciplines. We believe you will enjoy some absolute pearls of wisdom in their successes and failures as Firestarters.

DISCOVERING FIRE:
IGNITER, FUEL, AND ACCELERANT INTERACTION

Rub two sticks together long enough, and you'll get fire. It's all about the interaction of fuel (sticks), oxygen (air), and heat (friction). We believe the interaction of Igniter, Fuel, and Accelerant elements can light that spark in people to blaze into a Firestarter.

Researchers who study entrepreneurs have focused on many of the elements we consider part of the Firestarter Framework with a conceptualization of an entrepreneurial mindset. For example, in a comprehensive study, Mark Davis and his colleagues distinguished fourteen mindset dimensions, which tapped into Firestarter-relevant areas like freedom (independence), passion, mission focus (future focus), mastery (confidence), effort (persistence), and collaboration (interpersonal sensitivity).[17]

Firestarters each have strengths unique to their situations, but the one commonality is the interaction that occurs between elements. The largest fires are those that include one or more elements from each bucket. Igniters start the fire. Fuels feed the fire. Accelerants spread the fire. Successful Firestarters maximize their ability to tap as many elements as possible.

Additionally, becoming a Firestarter in any given discipline is a phenomenon that requires multiple aspects of the self and situation to be in harmony. Some people dismiss success as a by-product of personality. They believe some people achieve greatness because of whom they are and that there's nothing you can do to emulate that achievement. Others ascribe success purely to chance and simply wait for the moment to come—at some point in the distant future. It's simply not that simple.

The elements that we describe involve the interaction of the person, the situation, and the environment. Though intriguing, we do not discuss broader influences of culture and society, except tangentially. Collective and group-level elements are certainly important to the way Firestarters, and people in general, choose to behave and think. However, for us, the first step is isolating the individuals from overarching cultural influences to explore the aspects of their unique situations that have driven them to ignite their own lives and the lives of others.

HARNESSING IGNITERS

> *I am building a fire, and every day I train, I add more fuel. At just the right moment, I light the match.*
>
> —Mia Hamm, member of the
> National Soccer Hall of Fame

Often in cartoons, characters are displayed with thought bubbles that contain a light bulb. This is the universal symbol that an idea has been born. It's an *aha!* moment. Igniters are the triggers for the *aha!* moment waiting inside all of us. These triggers can include the following:

1. *Freedom in the moment.* Finding a situation that allows you the chance to determine your own path.
2. *Tapped talent.* Capitalizing on talent that you and those around you possess.
3. *Mastery mindset.* Confidence that you can be successful in a given skill domain.
4. *Driving passion.* Enjoyment, interest, and action driven by life's snapshots that grab and hold onto your attention.
5. *Global introspection.* Thinking strategically about yourself and how to maximize your productivity.

The act of igniting is a person-by-situation interaction. We believe that Firestarters likely have some common personality links, but the teachable value lies in combining the right person with the right situation. Firestarters are not distinguished by a desire to ignite everything they see. Rather, they are selective, choosing to focus their Igniters in ways that are most likely to have an impact. They are experts at finding and discerning the most appropriate situations to unleash their strengths.

Research psychologists have long recognized the importance of the person and situation working in tandem to produce results.[18] Firestarters succeed where others fail because they are better at assessing the potential of each situation and striking when they see the chance.

For example, successful Firestarters fan their passion continuously. This is illustrated in the following story about future Hall of Fame basketball player Kobe Bryant by a trainer who assists top basketball players with their shooting.

> One day I am at a gym with one of my clients, and Kobe Bryant is there. Bryant looks at my client with a harsh stare, says nothing to him, and walks off. I ask my client, "What is all that about?" and he has no idea. A little later Bryant looks at him with a harsh stare again and says, "I heard what you said." This occurs another time or two where Bryant says the exact same thing again. Finally, my client asks him point blank, "What are you talking about?" Bryant replies, "I heard you said [another player] was the toughest player to guard in the NBA. I'll be seeing you real soon."[19]

We did not independently verify the details of this story, but it nonetheless captures how Kobe Bryant continuously found situations that enabled him to ignite his passion. Even in his last game when the outcome wasn't going to impact the season, Bryant scored sixty points and led his team to a victory.[20] This is what Firestarters do.

FINDING FUELS

> *Do you know what my favorite renewable fuel is? An ecosystem for innovation.*
> —Thomas Friedman, *New York Times* columnist and Pulitzer Prize–winning author

As a kid, do you remember being fascinated by rockets taking flight? The countdown. You and your friends shouting, "Blast off!" The billowing smoke and flame seeming to engulf the bottom of the rocket. Then, just as it looks as though the rocket won't gain the power, it jets upward, leaving smoke and flame in its wake.

It's all about rocket fuel. We believe Fuels are critical for Firestarter success, and we define them as environmental resources that propel an idea, business, or life-changing choice. The Firestarter's environment could either be flush or lacking in the following Fuels:

1. *Prestige.* Social power due to hierarchical position, status, and control over outcomes.
2. *Opportunity.* An event, large or small, that leads to an increased possibility of success.
3. *Wealth.* The presence of your own or others' capital or assets that can be used to support action.
4. *Luck.* The conditions of probability that one can exploit for gains.
5. *Social connections.* A network of people available to be tapped for needed resources, information, or action.

All Fuels don't burn the same way. Prestige influences. Opportunity attracts. Wealth funds. Luck informs. Social connections supplement. As "burnable" resources, all Fuels have something in common. They have limits, and they can be squandered. You wouldn't burn a log on your fireplace when it's ninety-five degrees outside. In the same way, Firestarters have to be sparing with their Fuels and they must be right for the need in the situation.

Additionally, Fuels require Igniters to transform their state of energy. Gasoline does not auto-combust. A battery is useless until its positive and negative ends create a full circuit through some conduit. A Fuel without an Igniter is wasted potential.

ACTIVATING ACCELERANTS

> *There are no great limits to growth because there are no limits of human intelligence, imagination, and wonder.*
> —Ronald Reagan

Think about the effect of wind on a forest fire. Fires are alive; they consume and spread. The key to spreading is the presence of some action or force that drives the fire forward. For Firestarters, we call them Accelerants.

Oprah Winfrey began her career as a small-time chat show host in Baltimore, Maryland, in the 1970s. She rode the popularity of the program to a bigger hosting opportunity in Chicago, and her profile rose meteorically over the next several years for her to become the richest and most influential African American woman on the globe.[21] What happened? Oprah found a way to accelerate her success and exponentially increase her impact. She started a fire and didn't just watch it burn. She made it spread.

In the Firestarter Framework, we describe Accelerants as context-driven action. What does this mean? Acceleration happens because of something you do. It is behavior activated, and context is important. The same actions in different contexts are meaningless. For instance, partnering on trivial things doesn't matter. Partnering to make an impact does. Firestarters don't exponentially increase their impact by chance. They act in meaningful ways.

We highlight five specific elements under Accelerators in the Firestarter Framework.

1. *Mission focus.* Driven by a larger vision whether it is informed by experience, intuition, or faith.
2. *Collaboration.* Firestarters need other people who constitute the teams and partners needed to vet ideas or handle aspects of the work.
3. *Constructive competition.* A meaningful focus on competition as a motivator and idea generator.
4. *Sweat equity.* Investment of the raw effort needed for a fire to thrive.
5. *Support seeking.* Finding the mentors and experts to move forward in learning and growing.

Let's briefly explore how these elements aided Oprah's rise. When Oprah ended her long-running talk show in 2011, she could have walked away. She was in her midfifties and had enough money for ten lifetimes. She didn't take the money and run. She decided to embark on the massive expansion of a media empire through the Oprah Winfrey Network (OWN). She saw the potential for touching the world and broadening *her mission*, and she grabbed it.

Oprah had a team for all of her major endeavors. She *collaborated* with producers, directors, and business partners at every step. It is quite possible that some of those individuals were unhelpful or detracted from her progress. But a Firestarter learns to surround herself with those who prove themselves helpful to the cause. OWN was certainly not a solo endeavor. She had to identify the right team to drive the vision forward.

When Oprah began hosting *A.M. Chicago*, her ratings were low. Eventually, she led the show to beat her primary *competitor*, Phil Donahue, in the ratings by thousands of viewers.[22] The ratings were a valid indicator of the effectiveness of her position and her platform.

No one who looks at Oprah's résumé can say she didn't work hard. With a daily talk show, an acting career, and a production company to name a few endeavors, Oprah has invested huge amounts of *sweat equity*. She has achieved more success than most but has also exerted more effort consistently than anyone you can probably name.

When Oprah's talk show soared in popularity, she *sought support* by interacting with experts in multiple fields and shared her learning directly with her audience. Many of the experts whose knowledge helped her to grow were even able to ignite their own path through her show (i.e., psychologist Dr. Phil McGraw and fitness guru Bob Greene).

Oprah is a great example of Accelerants in action because her path clearly involved them all. Most Firestarters will not find every Accelerant to be necessary. The key is finding the right actions that combine with your unique Igniters and Fuels to set your growth in motion.

EXTINGUISHERS

> *Good fame is like fire; when you have kindled you may easily preserve it; but if you extinguish it, you will not easily kindle it again.*
>
> —Francis Bacon

Close your eyes and imagine a wisp of smoke, gently pluming toward the sky. This smoke could portend a future fire, but it could just as easily represent the smoldering remains of a fire that has been put out.

Why doesn't every fire, once kindled, survive and thrive? Why doesn't every person accomplish her dreams? The simple answer is that the Firestarter's life is a constant fight against extinction. Here are five elements that we consider Extinguishers:

1. *Discouragers.* Mental constructs like fear that douse your fire.
2. *Fuel limits.* Unsustainable nature of Fuels like wealth and social connections.
3. *Self-mismanagement.* Inability to manage the self (e.g., like in situations that involve gambling) or refusing to learn from mistakes.
4. *Punishers.* Consequences, fallout, and repercussions from actions.
5. *Ineptitude.* Lack of skill sets needed to achieve mission.

THE PHENOMENON OF COGNITIVE CONVERGENCE

> *When you look at people who are successful, you will find that*
> *they aren't the people who are motivated, but have consistency in*
> *their motivation.*
>
> —Arsene Wenger, French football manager

We all have people in our lives who seem to accomplish more in one workday than others accomplish in an entire week. Productivity follows those people regardless of what they are doing. PTA president. Nonprofit volunteering. Coordinating multiple kids' sporting events. Think about entrepreneurs who thrive on starting multiple businesses. The stress seems to propel them when it would destroy others. What explains this mountain of productivity? Why are some people able to be serial entrepreneurs when most would struggle with a single business? Firestarters form connections between seemingly disparate events in their lives, learn what works across situations, and are able to generalize this knowledge to other activities. We call this *cognitive convergence*, the ability to detect micro-signals in diverse situations that allow a Firestarter to ignite when others fizzle.

Cognitive convergence can be powerfully positive. Firestarters are able to make associations between similar situations and use lessons learned from one sphere of their lives to inform actions and thoughts in seemingly unrelated situations. They look for patterns of success, and then they pounce on situations that have proven to be generators of that success. Then they replicate that success.

This creates an overall sense of positivism and optimism. Firestarters reason that when problems arise, they will find a solution. They assume that the things they touch will work out—not because of pure chance but because of the strength of their actions.

Think about those in your life who approach problems from an optimistic perspective. Optimists tend to bring a "glass-half-full" mindset to multiple situations. This isn't an accident; it's a pattern. It works. When you believe positive things will happen, you find a way to make this vision a reality.

What others view as setbacks, Firestarters see as opportunities to learn and grow more. It goes beyond resilience. It is the reason why Donald Trump easily overcame obstacles in his presidential campaign that would have downed other candidates. It is also the reason why President Obama, even after eight years of embattled politics, authentically gave a final speech at the end of his term that was 100 percent as optimistic as when he campaigned in 2008.

It is the reason why Tom Brady believed his team was going to win when they were down 28-3 in Super Bowl LI. He was mentally locked into the idea that he could overcome a deficit that would have deflated others. It didn't matter to Brady that no other Super Bowl team had come back to win after trailing by more than ten points.[23] It wasn't important to him that there were less than two quarters left in the game. It didn't matter that it had so far been the worst game of the season. He didn't let self-doubt overwhelm him. Brady expected to win, and his teammates bought into it. As a result, the Patriots completed the most improbable comeback in the history of title games in professional sports.

Cognitive convergence is a learnable state of mind. There are, however, debates about just how to accomplish it. Learned industriousness theory and behavior-based psychology in general hold that any learnable category of performance can be strengthened by reward.[24] We all can imagine rats in cages repeatedly pressing levers to get food. In the behavioral psychology sense, we are all rats, pressing levers repeatedly in many areas of our lives. Despite being ensconced in psychological lore, some major opposing psychological theories (e.g., cognitive evaluation theory) hold that rewards can seem controlling and often undermine enjoyment and task choice for interesting activities.[25]

Which is it? Do rewards propel or thwart action? The reality may be more nuanced than simply saying rewards either help or hinder. Take creativity for example. Researchers examining tasks that require effort in the form of creativity find that it is increased by three specific reward conditions:[26]

1. Reward promised or expected for creative performance.
2. Reward offered for performance or another task following reward for creativity.
3. Action on a similar task after being rewarded for creativity.

These findings are consistent with the effects of reward on effort in other arenas (e.g., physical effort). From a learned industriousness perspective, Firestarters develop internal reward sensations to overcome aversion to the mental strain. Strain is something that people naturally find aversive. Firestarters overcome this aversion by seeing and anticipating the positive outcomes that result from the strain. Not only is the aversion to the strain reduced, but Firestarters seek it. It's an adrenaline rush that they often love and also works to their advantage.

Here are three ways to engineer your own cognitive convergence over time:

- First, affirm your vision. Free yourself to imagine the greater positive outcome that you are working toward. What will your fire allow you to accomplish? Momentary strain can be no match for a bright and positive vision of the future.
- Second, reward yourself for mission-relevant actions and results. When you achieve an outcome of significance, celebrate. Plan a timeline of celebrations that integrate with future opportunities that you perceive. When you capture an opportunity or find ways to increase your Fuel supply, take a moment for yourself and remember the benefit that the strain produced. Celebrating your success is refreshing. It creates an aura around you that others seek to replicate. It's the universal "I'll have what she's having" moment from *When Harry Met Sally.*
- Third, practice keeping your eyes open. Don't persist on one task blindly. Identify your other passions or things that interest you. As you get close to igniting other passions, note their similarity to other areas of your life where you are successful. Find ways to integrate your existing expertise and passion into areas that intrigue you but where you previously haven't allowed yourself to get off the sidelines.

Cognitive convergence also relates to the role identity perspective we discussed relating to the Innovator, Instigator, and Initiator types. People are motivated to maintain consistency between their public and private selves. Accordingly, we are constant observers of our own actions, tweaking and modifying as needed to ensure our action-based reality matches our internal identity. Since we have multiple identities, multiple acceptable actions align with whichever identities are most salient.[27] Someone who views herself as an Innovator will find a way to make creative interventions happen across situations.

Our actions converge with our overall perceptions. You are more likely to fly a plane if you identify as a pilot wannabe than if you identify as an acrophobic. Our actions also converge across situations. A lesson you learn as a pilot can be applied to other realms of your life. You may utilize experience with knobs and dashboards used to keep a plane aloft to help an entrepreneur friend design viable ways to measure her organization's success. This design would be "uniquely you." It represents a convergence of your experience across situations that few, if any, could replicate.

In summary, Firestarters don't let life dictate their actions. They've learned that action rewrites the script of life. They are industrious. And since research proves that industriousness can be learned, you can add positivism through cognitive convergence to your life by taking a more active role in that learning.[28]

EXPLORING EMISSIONS

> *A life is not important except in the impact it has on other lives.*
> —Jackie Robinson

When Jackie Robinson stepped on the baseball field in the 1940s, his actions created ripples that opened doors for African American and Latino players to be integrated in every major sport. It wasn't just his presence that would change sports and influence American society. It was the dignity in which he handled himself. This can best be summed up by this famous dialogue between Dodgers president Branch Rickey and Robinson:[29]

"Mr. Rickey, are you looking for a Negro who is afraid to fight back?"

"Robinson, I'm looking for a ballplayer with guts enough not to fight back!"

We call this kind of impact Emissions. Emissions occur when other fires can be started as a result of a fire that you've ignited and spread. Firestarters who spread their fire widely create platforms for Innovators, Instigators, and Initiators who they will never meet. Take a look at Steve Jobs's decision to launch Apple's App Store in July 2008.

Initially, Apple had eight hundred apps for its iOS operating system. By June 2016, there were over two million.[30] The large majority of those new apps were not developed by Apple, but they were enabled by its innovation. Or let's go back to our example of Elon Musk who has the potential to change the world with his every utterance. He can send a tweet about the need for tunnel boring in Los Angeles to alleviate traffic and people believe it will happen.[31] Why? Because his success has already created a platform that enables others.

Emissions may be completely unrelated to the original mission a Firestarter sets out to accomplish. In 1938, Roy Plunkett set out to develop a new type of coolant for refrigerators. The result was a heat-resistant product that was used in the Manhattan Project. Eventually, that material became a staple for cookware surfaces. It was Teflon.[32]

Emissions, in some ways, have a mind of their own. Once you set a fire in motion, you can't control where others end up helping it to spread. Firestarters often make an impact on the world in ways they never dreamed.

THE WAY FORWARD

We find the Firestarter Framework useful for assessing ourselves and the people who impact lives around us. We hope you see its utility and applicability to your life as well. Understanding Firestarters is important because these are the people who own the future. They are starting things, creating things, and disrupting things that will have ripple effects for the coming generations.

By understanding Firestarters, you become more aware of how to apply successful principles to your life. But that's not the only benefit. You will also understand the people around you who make an impact. You will discover ways that you can identify, support, and cultivate the Firestarters who inspire you and make a difference.

SPARK YOUR THINKING

1. What aspects of the Firestarter Framework resonate with you?
2. What parts of the Firestarter Framework are you most skeptical about?
3. Right at this moment, would you consider yourself an Innovator, Instigator, or Initiator?
4. While we have not addressed stages of life or age in this book, do you think that people are more likely to be Instigators or Innovators when they are younger? If so, does this affect the way you work with people of different generations?
5. Do you agree with our concept of "person-by-situation interaction"? Why?
6. Can you come up with your own examples that epitomize the Igniters, Fuels, Accelerants, and Extinguishers categories?
7. How can you improve your own life through cognitive convergence?

IGNITERS: WHAT LIGHTS THE FIRE

The Firestarter Framework

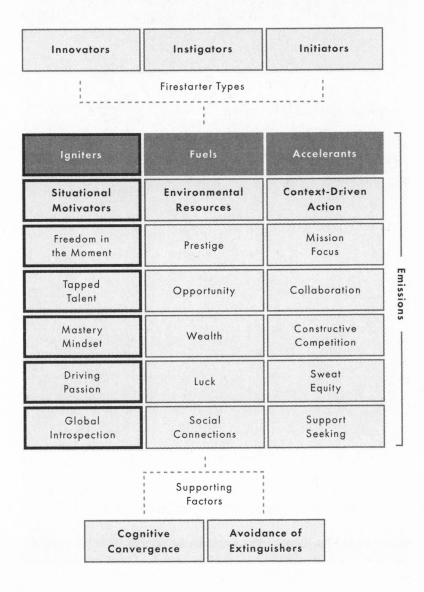

Innovators	Instigators	Initiators

Firestarter Types

Igniters	Fuels	Accelerants
Situational Motivators	Environmental Resources	Context-Driven Action
Freedom in the Moment	Prestige	Mission Focus
Tapped Talent	Opportunity	Collaboration
Mastery Mindset	Wealth	Constructive Competition
Driving Passion	Luck	Sweat Equity
Global Introspection	Social Connections	Support Seeking

Emissions

Supporting Factors

Cognitive Convergence	Avoidance of Extinguishers

Chapter 3

IGNITION POTENTIAL

W e are all born to be Igniters, but we haven't all found the situation that promotes ignition. Like car shoppers looking for the best deal, we are on a mission. We all dream about the perfect ride that can take us to our destination. When you get to the right place and see that car, you'll know it. But until then, you are in search mode.

We believe the *situational self* is the basis for ignition. What does this mean? You constantly flow between the tides of what you have just done, what you have been thinking about for days, what you're doing right now, and what you hope to do tomorrow. Your brain is a complex but limited tool. You can't think about the past, present, and future simultaneously. You can't daydream while paying close attention to reality.

Luckily, humans do have the capability to do all these things. They're just a little more iterative in nature. The situational self is not just represented by what you're physically experiencing in a particular moment. It's how you're interpreting and reacting to the moment in the context of everything else you're feeling, experiencing, and thinking.

Igniters are person-by-situation interactions. Something unique about the situation and something unique about the person click with each other. Often this click leads to the spark that Firestarters seek.

For years, social scientists and management professionals have been fascinated by intrinsic factors that drive interest, passion, and action. In his book *Drive*, Daniel Pink highlights a number of major research findings in the area of motivation and focuses on three intrinsic elements to motivation at work: autonomy, mastery, and purpose.[1] J. Robert Baum and Edwin Locke conducted a study of 229 architectural woodworking entrepreneurs.[2] They determined that the keys to venture growth over six years included passion, talent for finding and using resources, goal setting, and self-efficacy.

Our aim in examining Igniters is not to replicate Pink's or others' work. Rather, we aim to integrate our experiences, cultural anecdotes, and foundational research to describe an overarching worldview that incorporates the process of ignition as a necessary starting point to making an impact in the world.

The following figure shows the various person-by-situation interactions on which Firestarters capitalize to ignite their lives. Throughout this section, we explain these interactions in more detail.

Igniters: What Lights the Fire

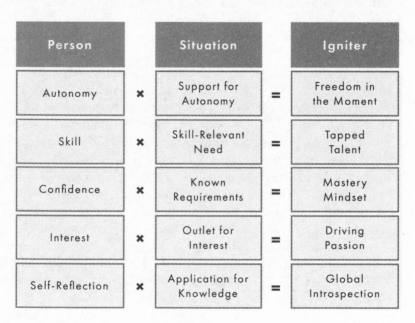

Person		Situation		Igniter
Autonomy	×	Support for Autonomy	=	Freedom in the Moment
Skill	×	Skill-Relevant Need	=	Tapped Talent
Confidence	×	Known Requirements	=	Mastery Mindset
Interest	×	Outlet for Interest	=	Driving Passion
Self-Reflection	×	Application for Knowledge	=	Global Introspection

Fig. 3.1. Igniters.

The key to igniting potential lies in recognizing situations that allows your freedom, talent, mastery mindset, passion, and introspection to play a role. The mere presence of an Igniter does not ensure a fire. Think about it. It's difficult (but not impossible) to start a campfire with a single match. But doesn't it seem impossible if the matchbox is empty? Igniters drive potential but are not a guarantee for success.

Not every situation rests on an "either/or" or "have it or don't" outlook. With a flint stone, the ignition potential depends on the strength of the strike, not just the presence of the stone. Every camper's dream is to have an entire box of matches to help start the campfire, rather than just one. With Firestarters, the principle is the same. More Igniters are better. Nonetheless, once you breach the threshold and start the fire, everything else is insurance against the wind.

There's an old adage that *success is* 10 percent inspiration and 90 percent perspiration. Some variations even have inspiration at a lowly 1 percent. What this adage doesn't tell you is that the perspiration can't happen without the inspiration. The 10 percent is the trigger, or the Igniter, that makes the perspiration meaningful.

SPARK YOUR THINKING

1. Do you believe that the situational self is the basis for ignition?
2. Looking at the list of Igniters, do any instinctively resonate with you?
3. Are there situations that are more likely to ignite you? What are they?
4. Is lack of one of these Igniters hindering your ability to achieve your dreams?

Chapter 4

FREEDOM IN THE MOMENT: LIVE ON YOUR TERMS

Freeing yourself was one thing; claiming ownership of that freed self was another.
—Toni Morrison, Pulitzer Prize–winning author

At the end of the movie *Braveheart*, Mel Gibson's character, William Wallace, is lying on his back preparing to be executed. A large crowd, including his fellow revolutionaries, has gathered to watch the final moments of their mentor, friend, and leader.

While writhing in pain, his captor informs the crowd that Wallace wishes to share some final words. The smugness of the guard insinuates that he doesn't think the injured Wallace has enough control to muster any words. Within seconds, Wallace unleashes a gut-wrenching cry of "Freedom!" that, for film effect, appears to echo across the countryside to be heard in the hearts of many people who have felt his impact.

"Freedom!" is the rallying cry of the Firestarter, as well. Firestarters celebrate the noble escape from the oppressed life of backward political regimes or parents who don't quite understand their calling. Freedom is both a struggle and a blessed light at the end of a tunnel of inspiration.

But what exactly is freedom? Freedom is a contextual phenomenon. You do not walk around constantly feeling free to do whatever you want, when you want, and how you want to do it. Every moment has its unique set of constraints and controls.

The drive for freedom swells in all of us. Firestarters are adept at pinpointing their spots to exploit that drive. Sometimes, it's as simple as taking initiative to develop the first draft of a document for a work team. Other times, you may have to escape a physical or figurative prison before you can act. Either way, Firestarters identify the factors of any given situation that support their drive for autonomy and take control.

SUPPORT FOR AUTONOMY VS. INDEPENDENCE

Freedom in any moment is a product of two things: the autonomy you feel and the support for autonomy that the moment allows. Often the terms "autonomy" and "independence" are deployed as synonyms for freedom. For Firestarters, however, there is a nuance. Autonomy is not the same as independence. One can imagine a teenager striking out on her own into the world without a friend. She has just escaped her controlling parents and stepped aboard a train to the great unknown without a penny in her pocket. She feels free for the first time in her life. She's finally independent.

Now imagine a similar teen who sets out to face the world with high support from her parents. She has many friends to call when times are hard. She has a bank account that is brimming with money from birthdays past. She feels free despite recognition that she sometimes still relies on her parents' support.

Both teens have freedom to do as they wish, but one is more likely to stumble upon success. Which one? The one whose autonomy is backed by parents and friends, right? Why is this? Both girls are equally free, but the second one has an added bonus. Her freedom is respected and supported by others. The first child may feel utterly hopeless and ineffective, despite her independence.

AUTONOMY AND THE SHADOW CHOICE

True freedom requires ambiguity. If a situation has an obvious choice of the "right thing to do," it gives an illusion of freedom but doesn't make you free. You technically have a choice of whether to pickpocket everyone you meet on the subway. The "choice" not to pickpocket probably doesn't enter the psyche of many people. Although free to do so, the obvious moral rightness of the choice negates any feeling that it is self-determined.

Moral imperatives are a bit murky in general. One could argue that society effectively reduces your ability to think freely about matters of something like theft from the moment of birth. One of the Firestarters featured in this book, don Miguel Ruiz Jr., comes from a discipline that calls this process "domestication."[1] The road to finding more personal freedom by examining and changing the beliefs or "agreements" you have made with yourself based on other people's and society's views is outlined in the classic self-help book by his father, *The Four Agreements*.

In the Firestarter Framework, freedom requires two steps:

1. Recognizing the presence of choice.
2. Having the ability to consider the alternatives.

If you don't stop and think about a course of action, you don't choose it. If the decision between alternatives is easy, you don't choose it. Choice requires a combination of contemplation and action. If you're not doing both, then you've convinced yourself you're making choices when you're not. On your deathbed, you will not think back about all the wallets you decided not to steal.

AUTONOMY AND THE MOMENT OF CONTROL

Some people believe that events happen to them. Others believe that they impact events. Researchers refer to this as one's locus of control.[2] The locus of control concept arose out of studies where researchers unexpectedly observed subjects who explained away (or changed expectations of) good or bad events by their perceived ability to make an impact on outcomes.

Imagine walking outside on a sunny day and opening an umbrella. All of a sudden, gray clouds sweep across the sky and the heavens burst open in a drenching rain. Would you think your action of opening the umbrella caused the rain to happen? Probably not.

Now imagine it is raining. You walk outside and open your umbrella. You arrive at your location dry. Another person walking next to you forgot his umbrella and arrives at his location soaked. Was your relative dryness within your control? Yes. You remembered to bring your umbrella.

Now here's a tough one. Was the other guy's wetness his fault? Your first reaction may be to say, "Yes, he forgot his umbrella." However, what you don't know is that the meteorologist he had watched on the morning news predicted a sunny day. In his mind, he had no control over his wetness because it was the meteorologist's fault. In your different minds, your dryness and his wetness had a different locus of control. You believe that you could control your dryness, but he believes he couldn't control his wetness.

Firestarters take the approach of personal responsibility for outcomes. Within every outcome, there is an element that was within your span of control. Sure, the meteorologist predicted sun, but why didn't the other person store a spare umbrella in his car, just in case?

If you peel back the layers of any outcome, there is always a point in any sequence of events at which your freely chosen action matters. Firestarters recognize the importance of locating this *moment of control* and taking action.

AUTONOMY AND SELF-DETERMINATION

The artist finds freedom through expression. The canvas speaks without words. Similarly, a poet drives feeling through words. She paints pictures without a brush. Someone who is instructed exactly what to paint or write won't *feel* the same way as someone who freely chose her direction.

There are two ways your action can feel self-determined. First, you can choose what to do. Second, you can choose how to do it. Constraints in either of these domains can lead to the feeling of being controlled. An artist directed what to paint loses some control. An artist directed how to paint it loses more.

As artists of success, Firestarters need expressive freedom for both the *what* and the *how* of their actions. When concepts aren't dreamed due to constraint, fear, or doubt, the world loses.

Edward Deci and Richard Ryan are two of the most prolific researchers on self-determination and its impact on an individual's intrinsic motivation. They have conducted in-depth investigations into how controlling aspects of the environment reduce motivation and how elements of the environment that promote self-determination enhance motivation.[3]

Deci and Ryan note that self-determination has been found to impact behavior across many domains including religious behavior, weight loss, adhering to medical regiments, political activity, environmental protection, and keeping New Year's resolutions.[4] People like freedom and respond favorably to it in a variety of circumstances. The key to feeling self-determined is to seek environments supportive of autonomy and avoid environments that institute controls or constraints.

SPARK YOUR THINKING

1. Do you allow yourself to be free, or do you accept the constraints that face you?
2. What beliefs do you have about yourself that keep you from igniting your potential?
3. What do you believe you control or do not control about your life? How do your beliefs about control affect the way you make decisions?

4. Do you have expressive freedom for both the *what* and the *how* of your actions?
5. What environments support your autonomy?
6. What environments do you need to avoid because they constrain or control you too much?

Chapter 5

TAPPED TALENT: USE THE GIFTS YOU HAVE

Talent must not be wasted. . . . [T]hose who have talent must hug it, embrace it, nurture it and share it lest it be taken away from you as fast as it was loaned to you.
—Frank Sinatra in an open letter to George Michael in *Los Angeles Times' Calendar*, September 16, 1990

Mozart died at age thirty-five but still managed to build a reputation as one of the most influential composers of all time. When we use the word "composer," he is likely one of the first names that drops in the collective consciousness, despite living in a time period with no social media. Why is this?

He had a natural skill set that spoke for itself. By age five, he was composing original music, not the typical modern kid's banging and strumming on Fisher Price toys. He toured European cities and went on to become an extremely prolific composer of operas, symphonies, concertos, and string quartets.[1]

Mozart represents an interesting case because he was raised by a father who quickly recognized and nurtured his talent. It certainly helped to live with someone who understood music and encouraged his unique abilities. How many potential prodigies in music and other fields go unnoticed or undiscovered because their talent is hidden or unknown to those around them? A key component of talent is being placed in a situation with a skill-relevant need where that talent shines.

RECOGNIZING SKILLS AND SKILL-RELEVANT NEEDS

Often in our roles as advisors, we hear people say, "I don't know what I'm good at" or "I want to do 'X,' but I don't know how." In response, we provide similar

49

versions of the following advice: "Everyone is a genius of their own experience. No one does you better than you." What does this mean?

You have a perspective. You have a combination of talents and influences. That combined perspective and group of talents is un-replicable. In his book *How to Fail at Everything and Still Win Big* and regularly on his dilbert.com blog, *Dilbert* creator Scott Adams has refined his description of an individualized talent stack or skill stack that individuals can tap to achieve great things in a way that is uniquely their own.[2]

When you enter a new situation, it's never really completely new, is it? There's always something that you've learned before that you can tap. Even language at its core is something with which you are familiar and use daily. Just recognizing the words that someone uses can help you make sense of an otherwise new world.

DOMAIN-RELEVANT SKILLS

Researchers sometimes refer to talent as "domain-relevant skills." This is one of the three contributors to creativity discussed by Amabile in her componential model of creativity that we discussed in chapter 2. What does this mean? For being creative or producing something innovative, talent matters. Innovation does not occur in a vacuum. Innovation does not occur by chance. Innovation is purposeful and driven by people who know what they're doing.

Now, there is a difference between talent and experience. We all know people who have performed a job their entire lives but who only rise to so-so levels. These people were unfortunately trapped in a context that did not support their natural talents. People who are talented have specialized knowledge that is either innate or learned. And they are able to put this knowledge into action. Action is necessary for something to be a talent.

Think about the sports commentators of a football broadcast. They may know the ins and outs of every play call and may be adept at spotting intricacies within formations. However, if you placed many of them on the field, many would crumble. They have the knowledge, but they don't always have the talent (even retired football players turned commentators may have dwindling returns on their talent). A certain threshold of knowledge is a necessary, but not sufficient, component of talent.

TALENT AND MOTIVATION

We discussed researchers Edward Deci and Richard Ryan in the last chapter. Their research has shown self-determination to be a necessary ingredient for intrinsic motivation. Their model also includes another ingredient: competence.[3] In brief, people enjoy doing, and they seek out opportunities to do things at which they're skilled.

Think about it. How many of you seek out opportunities to do things on which you always fail? Any takers? Why not? We don't like to fail. We like to succeed. We like to be good at things.

Accordingly, people pay attention to situational factors that let them know whether they are successful or not. Isn't this why we try to achieve personal best scores in video games and sporting events? We seek information that tells us how well we're doing.

As we have discussed, there is a lot of debate in the social science literature over the conditions when rewards can help or hinder motivation and performance. Deci and Ryan and other researchers have found that when rewards are verbal, conveying positive feedback about performance, motivation is enhanced.[4] Why is this? We like to be praised. Praise makes us feel how? Competent.

However, research on tangible rewards is fuzzier. In some circumstances, rewards can be perceived as controlling. A researcher trying to coerce action via reward would reduce a subject's perceived self-determination. However, even researchers like Deci and Ryan who adamantly oppose the controlling power of rewards recognize the potential informational value they can provide.

Reward procedures that convey competence would counteract some of the negative effects that would otherwise occur if people just felt controlled. When you achieve a standard or surpass someone else's performance level, the achievement serves as a satisfier of the need for competence and helps to motivate action.[5] To the extent that rewards convey this kind of competence, motivation may be maintained.

TAPPING OTHER PEOPLE'S TALENT FOR YOUR NEEDS

J. Robert Baum and Edwin Locke investigated a phenomenon that they called *new resource skill*.[6] Sometimes, a person who wants to start a business doesn't have all the talent or situational understanding or money needed to fund the venture. New resource skill provides an interesting explanation for why habitual entrepreneurs are often more reliable founders of successful businesses. Simply put, those who have done it once already are more likely to be successful than those who have never done it before.

A key aspect of the new resource skill involves identifying talented others who can help support the vision. You essentially tap the talents of others as a means to achieving your goals. Finding and relying on skilled others seems like a talent in itself, doesn't it? Success is not only dependent on understanding your own skill set. It's also important to recognize the talents of others and know how to profit from them.

TAPPED TALENT AND THE SITUATION

We have defined talent broadly to represent any developed skill set that one could use in a productive capacity. As with Mozart, talent could represent a natural inclination toward musical prowess. But talent could also be something intangible, such as an exceptional skill mooching off of friends.

Talent is situationally relevant. Musical skill would be beneficial when starting a rock band, whereas mooching would be beneficial if you lose your job and need money to scrape by (mooching and musicality are actually two very compatible talents for struggling musicians). For Mozart, his skill set was applicable to composing operas and masses, but he would have presumably fallen short in writing the next great novel.

Specific talents have limited and tangential situations in which they can be tapped with success. When the situation is right, it allows a Firestarter to shine. You have to be in tune with your own skill set. Are you more Mozart or Michael Jordan? Firestarters understand their varied competencies and are constantly on the lookout for broad applications. They are also adept at identifying the right time to rely on the skills of others.

SPARK YOUR THINKING

1. Do you have a skill that remains untapped? Is it purposeful?
2. Are you so consumed with your daily "busy work" that you forget to apply it?
3. Who can provide new resource skills to help you succeed?
4. What is your perspective on tangible rewards?

Chapter 6

MASTERY MINDSET:
BE CONFIDENT IN WHO YOU ARE

*We must have perseverance and above all confidence in our-
selves. We must believe that we are gifted for something and that
this thing must be attained.*

—Marie Curie

When Barack Obama first campaigned for president, you felt his
aura. Even when watching through the television screen, some-
thing was *there* that others didn't have. It emblazoned the zeal of his supporters
and annoyed those whose preferred candidates had to face him. As a little-
known senator in 2007, it seemed brazen to think that he could overcome
the might of the Clinton machine to become the Democratic nominee for
president.

But brazenness should be expected from someone who positioned the
uncommon term "audacity" in the title of his major memoir and drove it to the
center of common parlance.[1] Brazenness was on display when he delivered one
of his most effective campaign speeches following a *loss* in the New Hampshire primary. What should have been a celebratory night for Hillary Clinton
became the "Yes We Can" night for Obama.

Of course, people on the opposite side of the political spectrum likely saw
arrogance where supporters saw confidence. For the purposes of this book,
we distinguish between these terms in a key way. Confidence is an efficacious
attitude backed up by action; arrogance is the attitude in the absence of action.

By running for president, Barack Obama took action. By winning the
presidency, he proved that his confidence surpassed the requirements of myth.
You may have either liked or disliked his key policy agenda in the first years
of his presidency, but in the end he found ways to push it through a vocal
opposition.

We refer to this pervasive sense of confidence when you are aware of the situational requirements as possession of a mastery mindset. An individual with this way of thinking adopts a relentless vision of success and urges herself toward achievement.

MINDSET VS. TALENT

An argument could be made that a mastery mindset and talent are intertwining phenomena. After all, wouldn't people feel a lot more confident about activities where they are talented? On its face, this seems logical, but there are some key distinctions.

One of psychology's most influential theorists Albert Bandura has inspired a body of work on the concept of self-efficacy. He has shown that confidence and talent are separable concepts that differentially predict results.[2] Self-efficacy is a belief that you can successfully perform an action to achieve results in a given domain.

Bandura's research demonstrates that one's confidence, or self-efficacy, in a particular arena is a better predictor of results than actual ability. The outcomes of self-efficacy research over hundreds of studies provide two big takeaways:

- People often don't know their own talents (i.e., they lack confidence despite clear ability)
- Having a mastery mindset is more important than having actual mastery (even though mastery is often a prerequisite for complex tasks; for example, it would be difficult to be a professional quarterback or singer without actual mastery).

People's confidence in their abilities influences how they approach life.[3] Their dreams are likely anchored to what they feel they can achieve. They don't expend effort in areas where they expect limited internal reward. People with confidence will also persevere where others will give up. Even pure task interest is better predicted by self-efficacy than actual ability.[4]

Each situation has its own threshold for perceptions of mastery. When the talent threshold is minimal, confidence can be the main determinant of successful activity. No matter how hard you try, it would be difficult to build confidence in tasks pertaining to astrophysics, especially if you have no formal training. However, employees at a dry cleaner who have confidence that they can become the best pants pressers would likely outperform those who lack the same focus on mastery.

MASTERY AND FAILURE

People fail. It happens. Some children never win the biggest trophy. Some never get an A despite hours of studying. Eight out of ten businesses fail within eighteen months.[5] Those with high confidence ascribe failure to their own lack of effort, rather than lack of ability.[6] Such individuals are more likely to persist on tasks until success is achieved.[7]

Why are they able to succeed despite initial failures? Effort is fixable. Intelligence, talent, and personality flaws are harder to address. If you believe you have the ability, failure is an effort thing, not a talent thing. People with an effort-focused mindset persevere because they don't doubt the underlying ability to drive results. When the perceived deficit is effort-based, there is a solution—more or revamped effort.

After goals are achieved, those who are more confident raise their personal standards. Confidence leads to higher goals for the self. Period. Lack of confidence is a failure mindset. Failure that depresses confidence also depresses your future goals. Your belief in your mastery influences your choices, goals, effort, and perseverance.[8]

MASTERY AND ARROGANCE

As we described earlier, arrogance is confidence without action. Anyone can say, "I'm the best at this." But those with a true mastery mindset back up words with actions.

Some people attempt to project confidence as a social ploy. They want others to believe they are confident despite their own underlying doubts. Others are just lying to themselves. Here's an easy test in a given situation. When you feel confident, ask yourself why. Is it because you faced the same or a similar situation before? Have you seen a friend succeed? If you can't identify the reason why you're confident, you may just be bullshitting yourself or others.

So what types of experiences build perceptions of mastery? In his work on self-efficacy, Albert Bandura conducted several experiments on people with phobias of snakes. He investigated four methods for building real efficacy or confidence:[9]

1. A history of real experience with actual accomplishments. People who personally handle snakes and live to tell about it were more confident in future situations.

2. Vicarious experiences. Watching other models successfully handle snakes (or envisioning your own successful handling) increases your belief that you are able to do it.
3. Verbal persuasion. Participants are coaxed, encouraged, and informed how much their efforts are going to yield results.
4. Psychological state. Finding ways to remove the fear and arousal associated with snake handling allows people to isolate their debilitating emotions from the situation.

Real experiences with mastery are more impactful than imagined or vicarious ones—though vicarious experiences are more powerful than none. Additionally, verbal persuasion should not be viewed as a "pure" play. Simply telling you that you should be confident could backfire if your subsequent experience contradicts the persuasion.[10]

Imagine your friend has a pet snake and instructs you to not be afraid. You reach to pet the snake, and it bites you. No matter how much prior coaxing the friend provided, you're likely to be afraid of the snake in the future. You're also not likely to build greater trust in the friendship, are you?

Arrogance is a real risk when our prior experiences don't match our levels of confidence. Not all Firestarters are going to be expert snake handlers, but there is some advantage to those who have previously mastered the art of making an impact in a given domain.

MASTERY MINDSET AND THE SITUATION

People assess and integrate large amounts of information to judge their capabilities in various situations. Actual capability is only a part of this assessment. People are attracted to areas of perceived strength and allocate effort accordingly.

A mastery mindset is best achieved following training or experience that legitimizes a sense of confidence and efficacy.[11] Firestarters project confidence because of their ability to back up words with actions. They draw on lessons from their own and others' past experiences to find the motivation to tackle new or unknown adventures.

Mastery is a state of mind. It is powerful and can be initially tricked. You can convince yourself of your ability by self-persuasion, visualization, or repressing feelings of fear and arousal. But in the end, mastery is supported by action. Persuasion and visualization alone, without successful experiences, are only paper tigers.

SPARK YOUR THINKING

1. Do you believe that confidence is a better indicator of results than talent?
2. Are you comfortable thinking about yourself as a master?
3. What experiences have built your mastery mindset?
4. In what situations do you feel most confident and efficacious?
5. How can you find ways to replicate those experiences in other areas of your life?

Chapter 7

DRIVING PASSION: DO WHAT YOU LOVE

Every great dream begins with a dreamer. Always remember,
you have within you the strength, the patience, and the passion
to reach for the stars to change the world.

—Harriet Tubman

Some people love gardening. Others love baseball. Valentina Tereshkova loved plummeting to the earth from the sky and waiting until the last possible moment to pull the cord on her parachute. She performed parachute jumps more than ninety times (without her parents even knowing), sometimes planning water landings for the fun of it and sometimes at night to add an extra element of mystery.[1]

In 1963, at the height of the US-Soviet space race, Valentina became the first woman in space. She orbited the earth more times than any man ever had at the time—forty-eight. In addition, she broke the record for total flight time in space. She was only twenty-six years old.[2] How did she achieve this at such a young age? People had become aware of her skydiving passion, which drove her selection to be a cosmonaut.

Passion is an elusive phenomenon. When it is mentioned, most people can imagine it. Think about your own passions and how you feel when you are pursuing them. But how can you describe passion? Is it a burst of feeling or a calculated liking? Is it an attitude or an action?

In the literature on entrepreneurship, passion is a readily explored concept in itself. In social science research, investigators have found a very sterile way to explain passion using the phrase "intrinsic motivation" (or sometimes simply "intrinsic interest"). Intrinsic motivation is a multidimensional construct.[3] People could demonstrate an attitude-based interest in a topic, or they could behaviorally choose to perform an activity in their free time.

Passion is the reason for the presence or failure of many key situational interactions in our lives. People passionate about money seek opportunities for income. People passionate about being helpers find others to guide through life's travails. Not every situation drives passion in you, but every situation likely has someone who would be passionate about it if she sought it.

PASSION, INTEREST, AND EXCITEMENT

Think about a baby rattle or a jack-in-the-box. Are these things exciting to you? Could they mesmerize you all day? Probably not. Why is that? They're simple and easy to use. So easy a baby can use them.

This is why passion is tricky. Passion is complex. You aren't passionate about easy things. You're passionate about complex things you can be good at that others aren't. Things that tickle your mind in new and exciting ways.

A line of research known as flow theory discusses the importance of challenge for promoting engagement in tasks.[4] People in a state of flow are able to suppress goal-irrelevant distractions and become completely absorbed.[5] One key is reaching balance between too much and too little challenge. A baby's mind can be tickled by taking two steps forward. Yours can't. You need something bigger to drive you.

Melissa Cardon from Pace University and her colleagues set out to explore the concept of entrepreneurial passion as a driving force behind ventures and start-ups.[6] They described passion as "an intense positive emotion" related to entrepreneurial activities that motivates overcoming obstacles and promotes choice of activity in a given domain.

Cardon and her colleagues reviewed the history of the use of the word "passion" in entrepreneurial research. They found this commonality: research defines passion as involving "feelings that are hot, overpowering, and suffused with desire."[7] The use of the term "passion" is often intersected with words such as enthusiasm, zeal, and intense longing. Other researchers have found passion for work to be a core characteristic of wealth creation.[8] Passion helps entrepreneurs overcome extreme uncertainty and resource shortages.[9]

In other words, passion literally drives you forward in spite of obstacles. Think about those things that you thoroughly enjoy. Often you pursue them without even thinking about them, don't you? You recognize situations in which your passions can be pursued and drop other things just to do them. Interesting activities are self-conditioning.[10] You do things that are intrinsically interesting regardless of what reward or recommendation was offered by someone else. They are accompanied by a natural buzz in your mind. While a child may not want to clean his room, he may be interested in video games.

One of these activities would require a little more teeth pulling to accomplish, wouldn't it?

Interest is an attitude that precedes action. It is a holy grail for researchers because it is an individualized phenomenon. We're not all interested in the same things. Earlier, we described components of Deci and Ryan's cognitive evaluation theory that are proposed determinants of intrinsic motivation: self-determination (freedom) and competence (talent).

Intrinsic interest is also at the center of most major conceptualizations of creativity (e.g., Amabile's componential model). Accordingly, researchers have found intrinsic interest in one's job to lead to creativity among employees.[11]

Interest isn't just something you feel. It's something you live. It drives your obsessions. It drives your art. It drives your identity. Think about how you describe yourself to someone upon first meeting. Perhaps you discuss a musical genre that you enjoy, a sport that you play, or an aspect of your occupation that you have in common. You likely don't talk about all the things you hate or find boring. Interest doesn't just drive you; it drives how you interact with and relate to other people.

FREE-CHOICE ACTIVITY

What interests us and what we enjoy are the things that define us as people. People seek out clubs based on common political or social interests. Society is structured so that people self-organize into career fields based on their own passions. Interest drives action. Think about an activity that you don't care that much about. Your focus will more likely be on enduring or getting through it rather than worrying about specific nuances of the activity.

People are constantly bombarded with the hedonistic message that they should do what they enjoy. You can see it in advertising that involves young people freely experiencing the world. You can see it in common advice parents and mentors give to teens: *find a job that you enjoy and it won't feel like a job.*

Why is this enjoyment in action viewed to be so important? Passion and enjoyment are generally associated with positive aspects of mood, and research shows mood is used as an informational mechanism to inform when one should start or stop performing actions.[12] Behavioral scientists are able to induce either positive or negative moods in the laboratory. One such method is to place subjects into groups who watch either uplifting or sad movie clips. When given instructions to stop when the task is no longer enjoyable, those induced to have a positive mood persist longer at a task and produce greater output (such as naming types of birds) than those in negative moods. Those with positive moods also make quicker decisions when instructed to stop when they have "enough information" to make a decision.

In this context, the "do what you enjoy" instruction has some support. Firestarters naturally driven by enjoyment will persist longer, be more effective, and may make quicker decisions when working. People can tell when another person enjoys what she is doing. The most excited teacher is often a favorite among her students. The most energized waiter receives the biggest tips, and the happiest flight attendant creates the best flying experience.

PASSION, REWARDS, AND CONTROL

Praise is not controversial. Multiple meta-analyses (statistical reviews of other studies) have demonstrated the pervasive effects of positive feedback on task interest and free-choice pursuits.[13] Quite simply, we like to do what other people have patted us on the head for doing, especially when verbal reward signifies accomplishment. We think we're good at something when we're praised, and we want to continue to have our heads patted.

Tangible rewards offered for action are more of a mixed bag. Imagine this scenario. Two people are pursuing a dream. One arrived at the dream through her own self-discovery. The other had the dream foisted upon her through promises of pay or rewards and was directed to pursue it. One of these individuals is far more likely to ignite than the other. Meta-analyses show that tangible rewards, as external drivers of action, often undermine intrinsic motivation. Self-relevant passion is a stronger driver of action than external forces.

However, this negative effect is nuanced.[14] Unexpected rewards do not negatively impact interest and motivation. Additionally, under certain circumstances, rewards may still be beneficial. For example, rewards that provide information about competence (e.g., rewards for surpassing others' performance) may offer informational cues that outweigh any perceptions of being controlled.

Therefore, while controversial, external drivers may still be important contributors to action. Many of the Firestarters we talked to readily mention ego-driven goals such as fame, wealth, or family economic stability. There is no question that people like these things.

Where the distinction arises is how Firestarters integrate these external outcomes into their sense of self. Firestarters pursue their passions regardless of the economic or ego-relevant outcomes. External rewards are welcome by-products of the exploration of passion, but they are not the initial reasons for the exploration. They are ways to keep score and understand how successful their pursuit of passion has been.

PASSION AND THE SITUATION

In their study on entrepreneurial passion, Cardon and her colleagues suggest that passion can intersect with one's role identity. Remember, we highlighted three role identities that Firestarters may adopt: Innovators, Instigators, and Initiators. An Innovator is passionate about creating, an Instigator loves disrupting, and an Initiator loves to begin new journeys.

A powerful, exponential response to passion should result when a Firestarter's subject area of focus intersects with her role identity (especially the identity aligned with her default Firestarter type).

An Innovator is most passionate when doing creative things in the area where her passion has blossomed. An Instigator's passion would deepen when disrupting things when she has a high level of intrinsic interest in the subject. An Initiator would especially love starting things that align with the other passions in her life. Think about people who organize nonprofit charities. Many do so because a disease or other cause has directly impacted their lives.

The Firestarter role identities provide the perfect lens through which passion translates into action. Firestarters are self-aware. They know their passions. And they use this excitement to drive themselves and others toward goals.

SPARK YOUR THINKING

1. What passions drive you?
2. How do you ensure that your activities align with your passions?
3. Do you feel stuck in any activities that don't motivate you?
4. How can you flip the script and introduce your passions to the other activities in your life?
5. What external drivers such as fame, wealth, or family economic stability are important to you?

Chapter 8

GLOBAL INTROSPECTION:
CONQUER THE SELF

The first and best victory is to conquer self.

—Plato

Everyone isn't Brad Pitt. We don't wake up one morning and decide to purchase a Soviet-era tank as a reward for a hard day's work.[1] We're not iconic wrestler Goldberg, who bought himself a car as a reward for successfully coming back to World Wrestling Entertainment in 2016 after a thirteen-year hiatus.[2]

For most of us there's a little more struggle. Maybe you do something small. Like treating yourself to a favorite snack after "being good" on your diet for a week. Maybe you lock yourself in a room until you complete a blog post and reward your completion with a social media binge.

No matter your indulgence, there's a universal principle at play. People reflect. People react. People predict. The past, present, and future are intertwined. They live in the psyche simultaneously. Understanding how you reflect, react, and predict is the key to effective self-regulation.

In *Hamlet*, Polonius instructs his son Laertes, "This above all: to thine own self be true."[3] This is one of the quintessential struggles of human nature, isn't it? Learning what it means to be true to yourself. People do not act in a vacuum. They are connected to the world.

Accordingly, they must seek to understand their relationship to the world and seek to control it when possible. Global introspection is represented through the self-reflection activities you perform. How do you seek to understand yourself and your motivations? Introspection is a form of self-management. You reflect. You decide. You change. You allow yourself to grow.

INTROSPECTION AND GOALS

With strategic introspection, you control your own aspirations. You plan, act, and review performance, with the aim of objectivity. Objectivity requires that you have some way of knowing how well you're performing. You can't enhance your performance with praise you know is faulty. You can't lie to yourself. Part of you, deep in your cortex, always knows your true assessment of a situation.

Albert Bandura, who we discussed earlier, also inspired one of the seminal theories of self-regulation. He described the self-monitoring process as including three stages: self-observation, judgment, and self-evaluation.[4] These stages are driven by the temporal alignment of past, present, and future. One's present observations are driven by past evaluations, and together they drive future performance. If you know yourself and are attuned to your momentary thoughts, feelings, and reactions, you have greater control in terms of altering potential courses or setting new goals.

People who engage in strategic introspection think about goals in four dimensions. First, they attend to personal past information. They know how they've performed previously in similar situations. Research shows that people who aim to better themselves perform at a higher level than those who simply aim to match past performance.[5]

Second, they tune into momentary influences on action and performance. They adjust their expectations based on current influences. A placekicker with the wind at his back has much different expectations than one facing a 20-mph headwind. These types of momentary influences are why Scott Norwood is despised in Buffalo,[6] but Adam Vinatieri is loved in New England.[7] Norwood missed a field goal in the Super Bowl from a similar distance that Vinatieri nailed in the playoffs, now known as the most memorable play in Patriots history. The difference: Vinatieri made his field goal in seemingly impossible blizzard-like conditions.

Third, introspection involves integrating the expectations of others. What do your parents think? How about your friends? Do others believe in your abilities or doubt you?

Finally, people reflect on how others would likely perform on similar tasks. Your judgment of yourself, your abilities, and performance will vary widely depending with whom you choose to compare yourself. Playing a pickup basketball game against a team of two-year-olds would lead to vastly different expectations than playing against the 1992 USA Olympic Dream Team.

Research is very clear about the impact of two key aspects of goals on performance:

1. Explicit and challenging goals enhance motivation and achievement of outcomes.[8]
2. If there is no commitment to goals, they won't work.[9]

People who don't achieve their goals often become skeptical of their power. The key is self-reflection. You need to determine if the goal was the right goal. Was it challenging (or not challenging enough)? Did you consider your past performance appropriately? Did you compare yourself to the right targets (less than the Dream Team but greater than a two-year-old)?

Those who set goals for themselves outperform those who don't.[10] Those who challenge themselves on goals outperform those who satisfice. How do you know if you're challenged? Through appropriate, informed self-reflection. Change course when you misfire and reward yourself for approaching targets. People who institute procedures for self-reward will likely outperform those who do not incentivize themselves.[11]

Why do rewards matter for self-regulation but hinder other activities, as we described in chapter 7 on driving passion? Self-rewards as part of a true program of self-regulation are reinforcers of competence. You allow yourself to know, "I did it!"

Think of the writer who must generate a thousand words an hour to accomplish his book on time. Such a task may be crushingly stressful without a promise to self-celebrate each milestone. Here's the rub: reward only matters in introspective regulation systems if you really expect it when an outcome is achieved. If you continually withhold or forget your self-praise activities, they will come to have limited value.

INTROSPECTIVE REGULATION

Researchers James Gross and Oliver John identified a type of person they refer to as a reappraiser.[12] Reappraisers are in tune early with the actions and the emotions that are driving them. Being in tune allows them to quickly modify actions when needed. Those who reappraise are more likely to experience positive emotional states and behave accordingly, and they are less likely to experience and behave consistently with negative emotional states.

In a comprehensive set of studies, Gross and John found that reappraisers adopt optimistic attitudes and attempt to overcome negative emotions.[13] Reappraisers' friends like them more than non-appraisers, and they are wildly better off according to a grab bag of measures of well-being.

Introspection allows congruence between the self and the world. Why? Introspection allows you to trim the fat of ineffectiveness and inappropriateness so that perceptions of the rest of the world are consistent with the image you are trying to project. Introspection is a conscious way to subvert unconscious processes that would otherwise hinder you.

Think of the irresponsible employee who chooses not to attend meetings due to hair and nail appointments that occur while teleworking. It goes without saying that it's difficult for work and cosmetic appointments to occur simultaneously. Choices like this that go unchecked will eventually lead to incongruence between how you see yourself and how others perceive you. You won't even know how the incongruence arose if you aren't being introspective. Self-understanding allows the process to be set in motion for corrective change.[14]

INTROSPECTION AND TRUTH

Imagine a dartboard with a large bull's-eye that repeatedly fades away and then reappears. You throw a dart and hit an area where the bull's-eye once presented itself but doesn't anymore. Do you give yourself the points? Would your answer be different if you were playing against yourself or in a competition with others?

Self-observation only matters if what you're observing is accompanied by clear evidence. Ambiguity provides room for misinterpretation. Ambiguity allows you to lie to yourself. When developing a model of self-regulation, people must decide what models to compare themselves against and what measures to use. Poor choices in either of these dimensions will yield inaccurate comparisons, and you may not even be aware that the comparison is wrong.

Sources of information for introspective goals can be one's prior performance level, one's own standards, the performance of others, or others' expectations. Your judgment of the same performance in yourself will vary widely based on the object of your comparison.

Introspection allows you to determine if success (by any of these measures) was something driven by your activity or whether other factors were involved. Do not just blindly accept (or perpetuate the myth) that anything great you accomplish was your doing—and failures were caused by something else. At the same time, don't discount your successes in the face of minor failures. For a long time, researchers have understood the power of attending to success versus failure of your actions.[15] Attention to successes improves performance. A pure focus on failure will likely lower future performance.

Similarly, lying to yourself is a failure's strategy. If you lie too often, you will misunderstand the reasons behind your own past actions and results. Your frame of reference becomes skewed, and any lessons you attempt to integrate become tethered to a foundation of quicksand. The future is motivating if you can anticipate it. The way to predict accurately is to reflect on patterns of the past, test hypotheses, and adjust your worldview as you collect new data in the present.

INTROSPECTION AND THE SITUATION

Many of the Firestarters we interviewed mentioned external motivators for success, such as fame, money, and security. Now, many people experience these motivators, but most don't come close to being Firestarters, do they?

A key difference is what is known as *controlled aspirations*. Firestarters find ways within the situation to convert external feedback systems into internally applicable standards. They don't just desire money. They develop an internal system for continuously evaluating their level of wealth versus the desired state, root causes for the discrepancy, and systematically choosing courses of action that close the gap. They don't just want fame. They find ways to exercise control over the processes that are likely to drive them to that end.

Introspection matters to results when your goals in a particular area are clear and your self-evaluations are honest. The act of introspection strengthens your environmental recognition and your cognitive convergence activity. When you reflect, you recognize a situation that resembles one that led to past learning, and you are able to pounce by applying your new understanding of how to be most successful.

SPARK YOUR THINKING

1. How often do you pay attention to why you do what you do in a truly introspective way?
2. How does this feel different from your default approach to life?
3. Do you agree that ambiguity allows you to lie to yourself? Why?
4. How do you control your aspirations?

FUEL: WHAT FEEDS THE FIRE

The Firestarter Framework

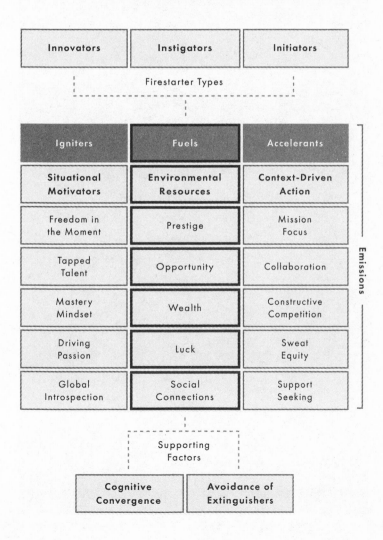

Chapter 9

FUELING THE FIRE

A standard venture capitalist question goes something like this: would you rather be rich or be the king? The aim is to determine whether a person is a founder or an entrepreneur. Should the founder continue leading the organization or cash out when the organization is strong enough to live on its own?[1]

From a Firestarter perspective, being rich and being the king are both helpful scenarios. Wealth and prestige are resources that can be tapped to further your goals and make an impact. Organizational research and practice often use resource-related language to describe elements of the environment necessary for optimal functioning. Think of the term "human resources."

In the Firestarter Framework, we describe Fuels as environmental resources. We distinguish between two environments that can be tapped—the one that revolves around the self and the one that revolves around the situation.

Fuels are resources for use in achieving a higher purpose. Successful Firestarters gain understanding of how to find and mine these resources. In the figure below, we show the resources that exist in the *environment of the self* and the *environment of the situation*.

The environment of the self contains resources that are due largely to you being who you are. Whereas the Igniters discussed in the prior section mostly represent internal mental motivators, resources in the environment of the self are external to the mind.

The environment of the situation contains resources that are due largely to the circumstance in which you find yourself. These resources exist because of people you know and how others perceive you.

Throughout this section, we explain each of these Fuels in more detail.

Fuels: What Feeds the Fire

	Environment of the Self	Environment of the Situation
Prestige	Position	Power
Opportunity	Planned Moments	Happenstance
Wealth	Bank Account	Other People's Capital
Luck	Calculated Risk	Probability
Social Connections	Personal Rolodex	Networking Potential

Fig. 9.1. Fuels.

Successful entrepreneurship involves acquisition, combination, and redeployment of resources.[2] Similarly, being a successful Firestarter involves acquisition, combination, and usage of Fuels. You can't sustain a fire without something to burn.

Fuels are everywhere, and often any barrier to using them is more psychological than physical. How do you use the prestige, opportunity, wealth, luck, and social connections in your life to power your ideas and make an impact? Don't discount this question too quickly.

Resources are available in the environment of the self and the environment of the situation. People often claim that they can't start a fire because they're not rich enough, they're not lucky, or they don't know the right people. This is excuse making, not reasoning.

If you feel you don't have prestige, think about how you've tried to structure situations for maximum control over yourself. If you haven't *happened upon* the right opportunities, think about whether you've done any active planning to make those activities come to you.

If you don't have money or prefer to keep your funds in savings, ask yourself how you've tried to utilize others' capital. If you don't feel lucky, think about the times you formally sat down to determine the probability of success in different situations. If you feel you don't know the right people, turn to the next person you see on the street and introduce yourself—that connection could be the key to your future.

As we described the actions above, did any of them make you feel uncomfortable? Probably. Why? Because you know there are situations where you had resources available to burn, and you blew it. You didn't use them—not because the resources don't exist. You blocked yourself from using them.

Many of the Firestarters we feature in this book have strong views on resources. Just listen to seventy-six-year-old Jerrie Ueberle: "There is nothing that doesn't have resources, answers, or access to answers. We live in a time of great resources, and that we have the poverty and lack of connectivity is shameful."[3]

SPARK YOUR THINKING

1. From our list of Fuels that feed the fire, which ones are most applicable to you and why?
2. Do you have an example in your life where you can differentiate between an environment of the self and an environment of the situation?
3. What are the barriers that are preventing you from using Fuels?

Chapter 10

PRESTIGE: DELIVER ON YOUR PROMISE

With great power comes great responsibility.
—Uncle Ben, from the 2002 film *Spider-Man*

I n 1961, Robert Kennedy became attorney general in his brother's presidential administration at age thirty-five. Was he selected after an exhaustive search? Many in the Senate and around the country didn't think so. An argument could be made that he had impressive qualifications, but his ascension and Senate approval were ultimately aided by his brother's office.

Despite the appearance of nepotism and public doubt, Kennedy embraced his position to become one of the most influential members of the cabinet, a major contributor to the civil rights movement, and a devoted, effective opponent of organized crime. Upon his brother's untimely death, Kennedy sought and won a seat as senator from New York and eventually ran for president on his own laurels before his unfortunate assassination.

Kennedy had position, name, and access to power that few will know. Yet many people with status or power choose to sit back and relax or live a private life. These decisions can be respected in a world of free choice, but there is something inspiring about someone who strives to make an impact when her circumstances don't require it. It's a good news story for society when the prestige of position and power are used as Fuels to power something greater.

Prestige is a societal construct. Prestige represents positive prejudicial views that others have about your efficacy. This contrasts with how we described mastery mindset, which pertains to the confidence that you perceive in yourself. Prestige arises due to your hierarchical position, membership in a prized social group, or direct power over others' outcomes. The corresponding perceptions others maintain about you are powerful sources of energy. Firestarters tap them and use them in ways that promote their desired area of impact.

THE PRESTIGE OF POSITION AND STATUS

Firestarters love starting their own companies. Why? Because entrepreneurs have instant prestige. In many situations, a business card with the title "CEO" can pack more punch than the title "director" for a multinational corporation, even if the director earns ten times the income. When you become an entrepreneur, you create power and prestige out of thin air. It's like when the Federal Reserve creates new money, but the currency inside you is prestige and status. You have more internal capital simply by signing your name on a legal form.

Think back to the venture capitalist question at the beginning of the last chapter. Someone who wants to be king is more likely to use prestige as a Fuel. Why? Because when you're king, you have prestige built into your role. It's accessible, and everyone recognizes it.

Similarly, some people belong to social groups (either chosen or demographic) that immediately convey a level of status. Membership in groups matter for the resources they provide to fuel your progress.

If someone says he or she is a member of Mensa, you form a perception, don't you? Someone on the board of trustees of a thriving nonprofit conveys certain characteristics and values, even if you don't know the person. Your status fluctuates wildly from one situation to another, but it is always due to your relative position in a group. Think about a mid-level manager in an organization. The prestige you hold in the organization that pays you may be great, but if you walk into the doors of another company, no one knows you and your position means less. As your environment changes, your status changes.

PRESTIGE AND POWER

Researchers exploring social interdependence view power as orienting from social games that everyone plays. According to interdependence theory, any time you interact with another person or group to achieve a joint outcome, there is a power structure to the situation. John Thibaut and Harold Kelley identified three types of control underlying interdependent situations:[1]

1. Control over your own outcomes (often called reflexive control).
2. Others' control over your outcomes or your control over others' outcomes (often called fate control).
3. Joint control over each other's outcomes (often called interdependence).

Obviously you would enjoy maintaining reflexive control when possible. There is a certain level of stress removed from situations when the outcome is within your direct control. You would also prefer situations where you maintain fate control over others, rather than the opposite (unless you had extreme trust in the other person).

Interdependent scenarios would be more stressful because the achievable outcomes are less predictable. What you can achieve is literally dependent on what others decide to do within the context of your own choice. An equal action by a partner could have positive or negative results depending on your own course of action. Similarly, an equivalent action by you could have positive or negative results depending on what your partner chooses. In interdependent situations, both you and someone else have power, and the ambiguity can be frightening.

Your tap-able power would differ depending on the type of control you hold in any situation. When you control your own outcomes, you can move relatively quickly. Your fate is literally in your own hands. Firestarters examine the environment for opportunities to exercise reflexive control. This type of power represents the "low-hanging fruit" of impact.

But is reflexive control really a form of power? You may admire someone who effectively loses weight through a strict diet. But the power is somewhat isolated. When you look at someone "doing her own thing," you don't necessarily attribute her actions to power. Why? Because the power present in those situations is not socially relevant.

When you control others' outcomes, they are more likely to recognize your power. Accordingly, scenarios that involve interdependence or fate control over others are more prestigious. Your power is both real and socially palpable.

Although a bit scarier, Firestarters prefer the power inherent in interdependent relationships versus those in which they only have fate control. Why is this? In interdependent relationships, you still have control over your own outcomes. With fate control, you only control the outcomes of others.

Think about a dog going through obedience training. You learn to control his behavior through treats and praise. You literally control his fate. When he performs a trick as desired, you give him a cookie. However, when he performs differently than desired, you withhold the treat. He may give you those puppy dog eyes and make you feel guilty. But you can't give in.

When you rely on fate control, you can end up hurting others without achieving your desired results. You don't take a dog to obedience school for the fun of it. It has a purpose. If the dog doesn't respond to the treats (the power that you have), then your ability to capitalize on the control is limited.

Firestarters look to capitalize on interdependence. Why? It allows them to negotiate. It allows them to persuade. Firestarters need to frame win-win scenarios in ways that appeal to the others who benefit from the joint outcome. Firestarters seek ways to avoid being subjected to others who have too much fate control over their outcomes, and they find ways to maximize results when outcomes are interdependent.

PRESTIGE AS A RESOURCE

Having prestige is not an end in itself for Firestarters. It must be used to further higher goals.

Think about Congress. What's the difference between those representatives and senators who take action and those who sit back? Those who move are tapping the resource of prestige and using it. Those who sit back are letting an important resource lie dormant. It's the same as buying a ranch with stores of petroleum and resisting the urge to drill. They know it's there but don't seek to use it.

Prestige is social capital. It exists in all situations in which at least two people are interacting. Sometimes it is more valuable than others, like when it is wielded by a king or CEO. But most non-kings and non-CEOs face situations every day when it can be tapped. The key is attuning to it and recognizing when the power you have can be used in any situation to support your goals.

SPARK YOUR THINKING

1. What examples of reflexive control do you have in your life?
2. In what areas of your life do you have fate control?
3. Is interdependence a state you strive for in relationships? Why or why not?
4. How does your level of prestige affect your interactions with others?
5. Is status important to you? When does it help you, and when does it hinder you?

Chapter 11

OPPORTUNITY: OWN YOUR MOMENT

You have to come to your closed doors before you get to your open doors. . . . What if you knew you had to go through 32 closed doors before you got to your open door? Well, then you'd come to closed door number eight and you'd think, "Great, I got another one out of the way." . . . Keep moving forward.
—Joel Osteen, pastor of Lakewood Church

For fifteen seasons (with more now in the queue), the television show *American Idol* racked up ratings partially on the backs of young people with no singing talent. The first several shows of the season were dedicated to bloopers of people's ego-driven, fame-seeking exploits and out-of-tune caterwauling. Talented singers were sprinkled in here or there for good measure.

William Hung is an example of someone whose *talent* would usually never be displayed during a singing competition. With a somewhat nerdy appearance, he seemed comically unaware of his own lack of singing ability. His rendition of "She Bangs" was gut-wrenchingly bad but delivered with such confidence that it didn't matter. He became an immediate sensation and even went on tours showing off his "abilities."

Viewers and the show's judges laughed at William Hung. But there was a certain irony in his ascent. Much better singers were turned away. Fantastic singers stayed at home and never auditioned. Hung, despite a lack of traditional talent, achieved the dream that many others never would (at least the fifteen seconds of fame variety). Why is that? He auditioned. He embraced his uniqueness. He sought opportunity where others sat on their couches.

WHEN OPPORTUNITY DOESN'T KNOCK, YOU SEEK IT

Do you want to know a secret? Many opportunities are planned ahead of time. They don't just happen. They are sought. Some people go to college—why? To learn. Sure. But also to open themselves to opportunity. Everything you do in life prepares you to recognize new opportunities. Think of Kathy Ireland who started as a supermodel but who became an extremely successful business-woman with a worldwide brand.

The first step of her business enterprise outside of modeling was developing her own brand of socks, and now her company has more than fifteen thousand products.[1] She found an opportunity to bundle her model persona with a business that sells clothing, furniture, and jewelry. This didn't just happen to her. Think of all the models in the world who never start multimillion-dollar businesses. The business was, for her, a natural expansion of her persona. It also capitalized on skill sets she possessed that other models don't. She created her own opportunity by recognizing her unique potential for brand expansion.

For many, this planned version of opportunity is more comfortable. Think about the thousands of people every year who meticulously craft business plans to capitalize on their venture ideas. Business plans aren't necessary, but they are great for convincing others (including venture capitalists) to join you on your wild journey. The opportunity exists. The business plan formalizes it.

The key to capturing opportunity is time. Time allows you to think. Time allows you to plan. Time is often considered a resource in itself, but the true importance of time lies in the opportunities it permits you to pursue when you have it.

SOME OPPORTUNITIES JUST HAPPEN

Yogi Berra once said in his classic way, "When you come to a fork in the road, take it."[2] If you think about this quote too long, it ties your mind in knots. Why? Because you can't take both roads, can you? Consider this. No matter which road you choose, opportunities await that you never dreamed or anticipated.

When Jack Ma searched the internet for the word "beer" in 1995, he found no information about Chinese brews. Nor did he find much information about China in general. He saw an opening and started his first website about China. Years later, he was executive chairman of one of the largest internet companies in the world, Alibaba Group. And it all traces back to recognizing a gap in the world of beer![3]

Would Ma have been successful if "beer" had returned dozens of Chinese search results? Possibly. But this was the opportunity he saw in that moment. He hadn't necessarily dreamed of being an internet entrepreneur. In fact, he had just recently heard of the internet.[4] Although planning can be helpful, recognizing opportunity doesn't take years of planning. It often involves simple observation and a moment of insight.

When relying on opportunity in the moment, you need quick reflexes. Most people feel more comfortable with the planned type of opportunity. Why? You feel more in control. With fleeting opportunities, time is literally against you. You see the opportunity and you have to move or not. You don't have time to weigh the pros and cons or to develop a SWOT analysis. You just need to decide quickly whether to take the plunge.

OPPORTUNITY AS A RESOURCE

Opportunities are only as good as the Igniters you deploy to take advantage of them. Individuals with lack of mastery mindset are less likely to embrace an opportunity even if they could act on it and drive personal success. Also, if you have a limited sense of freedom in a given situation, you may not feel that an opportunity is driven by your choice.

Belief that you can act is a powerful motivator. Belief that change can happen *in a flash* is an even stronger motivator. Opportunity exists around us in all capacities, so many in fact that you could never possibly take all the opportunities available to you.

Best-selling author Steve Pemberton didn't feel as if he had much of a choice as a child. He grew up in a foster care system that shuffled him around. Eventually a kind neighbor helped him realize that there were people who could care for him, and this opened his heart to a new world of possibilities. Pemberton went on to become a chief diversity officer for Monster.com, the first chief diversity officer in Walgreens history, and a well-known speaker and author who has inspired millions with a book and film based on of his life. When we spoke with Pemberton, he mentioned that he approaches decision making from a perspective of unlimited possibilities. Accordingly, he almost never feels deterred from a big idea.[5]

As a resource, opportunity is tricky. When it is a planned resource, you could spend years cultivating just the right set of knowledge and experience to be able to pounce when the right moment comes. But for others, opportunity is a flash in the pan. Think of the music producer who just happens to hear an undiscovered voice at a hole-in-the-wall pub. For the musician, the moment is more planned. She's been gigging for years hoping for such a discovery. They both experience the same opportunity with different levels of foresight.

Because there are so many opportunities available, you will always have doubt. Is there something greater that you could have achieved if you had waited for or worked toward a different opportunity? In the end, though, questioning yourself and the situation repeatedly will only ensure that no action is taken. Without action, all opportunities are wasted resources.

SPARK YOUR THINKING

1. What opportunities have you not acted upon that you now regret?
2. If you replay the scenario in your head, what would you do differently?
3. Are you more comfortable with planned or unplanned opportunities?
4. What planned opportunity can you work on this year?

Chapter 12

WEALTH: PAY IT FORWARD

The use of money is all the advantage there is in having it.
—Benjamin Franklin

In the *Justice League* movie trailer, the marketers highlight a particular joke from the movie that is funny because it is true. The Flash asks Bruce Wayne (Batman), "What are your super powers again?" In reply, Wayne quips, "I'm rich."[1]

We all agree: having your own wealth is great. But sometimes, you need to tap the resources of others. In the book *Les Miserables*, Jean Valjean starts life anew as a penniless, released prisoner. He seeks asylum for the night under the roof of a generous and pious bishop. Valjean repays the bishop by absconding with some valuable silverware. When police nab Valjean, they attempt to return the stolen items to the bishop.

Instead of corroborating the theft, the bishop offers Valjean some additional silver candlesticks. This moment was pivotal in Valjean's transition from a vagrant to a savvy businessman.

We are in no ways advocating theft from the clergy. This example highlights the importance of finding sources of wealth (often other than one's own wallet) in helping to fuel the fire. Wealth is more than what you have in your pocket or what is in your bank account. It also exists in what you can *legally* access from family, benefactors, investors, or the bank. Firestarters don't need personal wealth to use it as a Fuel (though it certainly helps). They just need access to it.

WEALTH AND YOUR BANK ACCOUNT

In the famous song "If I Were a Rich Man" from *Fiddler on the Roof*, Tevye yearns to be rich so he won't have to work so hard. Often, this idea is held up as an ideal. If you get rich enough, you'll never have to work again. If you win the lottery, it's the golden ticket to sipping margaritas on a beach for eternity.

Firestarters don't think this way. Having wealth (even a small amount of wealth) is not the end game for them; it's a tap-able resource for fueling something bigger. Think of the entrepreneur who drains his bank account for a dream. If he fails, his family will lose home and hearth. Is he stupid? Is he reckless? Perhaps, but that's beside the point. Firestarters find ways to put wealth into action.

Driven by a need to make an impact, they are constantly seeking ways to make their money an equal partner in the journey. Can they invest in income-producing assets? Can they buy a new tool or item that others need and rent it out? Can they purchase source materials and fund a manufacturing run for a new invention?

The idea that money begets more money and success is not rocket science. Yet, the world has plenty of rich people sipping margaritas. Are they bad people? Of course not. They're allowed to use their money how they please.

The world is also full of people with modest bank accounts who choose to sit on it for safety and security reasons. Are they incompetent? Of course not. Our purpose is not to cast aspersions on those who choose passive approaches to their wealth. Rather, we aim to make a distinction. The makeup of a Firestarter does not allow for a lifetime of margarita sipping or passive waiting to achieve a level of security. There's an itch or a drive that rises to the surface and requires action. Access to wealth of any amount simply allows a Firestarter to scratch that itch.

WEALTH AND OTHER PEOPLE

You don't have to fund your own way to use wealth as a Fuel. Think about the concept of a mortgage. Most people who own a home have a mortgage. Why? They don't have a few hundred thousand dollars lying around. The mortgage allows you to enjoy the benefits of homeownership while paying a smaller monthly payment to a bank.

Similarly, businesses leverage mortgages and other loans to maintain a cash supply and cash flow. In his Rich Dad series of books and educational materials, Robert Kiyosaki often extols the virtue of using other people's money (OPM) to fund your success.[2] Banks, venture capitalists, and investors look to lend money and fund ventures not because their hearts are golden but because they can't accomplish their cash flow missions on their own. Their results are bundled within others' projects.

In recent years, access to other people's capital has been democratized through crowd-funding platforms like Kickstarter.com, Indiegogo.com, GoFundMe.com, and Patreon.com. Nonprofit organizations also get in on the action through sites such as Crowdrise.com. Anyone who has an idea can post prototypes, examples, descriptions, and explanations and immediately start using the web as a tool to fund her inspiration. Firestarters with passion do not have monetary excuses for not moving forward with projects.

Kickstarter maintains stats of successful funding efforts. As of September 2017, over 35 percent of projects (over 125,000) had been successfully funded.[3] While most successful projects are under $10,000, there are many examples of six-, seven-, and even eight-figure projects that took off. The top funded projects include watches, games, music players, and films.[4]

Behind all of these platforms, there is often an old-fashioned source: family and friends. People close to Firestarters are often the ones who fund their efforts, either online or in other ways. Part of the reason why web-based crowd-funding platforms are popular is because they formalize (and impersonalize) a process that once had to be accomplished informally and face-to-face. It's a lot easier to post a link to a crowd-funding platform on Facebook than to approach someone in person and ask for money.

Getting the funds to nourish your impact has never been easier. With the current suite of tools available to anyone who has internet access, lack of money is often a convenient excuse for lack of effort.

WEALTH AS A RESOURCE

As far as resources go, wealth seems the most tangible. It's real and spendable. Some people inherit great amounts of money and use it to change the world. Others blow all their money or live paycheck to paycheck. Others choose to play it safe with their money and invest for the long haul.

You probably often think about the things you could do if you only "had the money." Even if you do have some money, you still think about this. Even Jeff Bezos, founder of Amazon, probably thinks about this. Whatever your situation in life, "having the money" is simply a proxy for thinking about the slightly richer version of yourself.

As a tangible resource, wealth is also very wastable. While Firestarters are driven to use wealth in ways that others aren't, they must also be careful to ensure that the drive doesn't push them off the financial cliff. Wealth can be used, but it also must be nurtured so that it is available as a continued resource and not just a one-off binge.

SPARK YOUR THINKING

1. What is your personal mindset when it comes to money?
2. Are you comfortable approaching friends and family to invest in your ventures? Why or why not?
3. If you examine your spending habits over the last five years, how much of your financial resources were reinvested in achieving your dreams?
4. Do you tend to admire Firestarters who started from wealth more than those who started from nothing? Why?
5. Do you believe crowd funding is a viable way of raising money to ignite an idea you have?
6. How would you figure out what financial resources you need to accomplish your vision?

Chapter 13

LUCK: PLAY THE ODDS

If you're playing a poker game and you look around the table and can't tell who the sucker is, it's you.

—Paul Newman

In 1991, Dr. Nicholas Terrett filed for a patent for sildenafil citrate to serve as a heart medication.[1] Within a few years, this medication became one of the most popular drugs under Pfizer's brand. But it wasn't being used as a heart medication. It was sold as Viagra, the little blue bill used to treat erectile dysfunction. How did this happen?

During clinical trials, treatment subjects reported increased incidence of erections, even several days later. The new use for the drug was discovered through analyses involving statistics and probability. Scientists use probability-based luck (wrapped in the scientific method) as tools for uncovering the secrets of the world.

Firestarters with a keen mathematical and logical sense have the ability to play these odds better than others. But not all luck is wrapped in a controlled scientific experiment.

We usually think of luck as intertwined with the concept of chance and blindly stumbling upon success. But Firestarters find safety in luck. Why is this? Most people view luck as uncontrollable. Firestarters view events with a component of luck as decision points. You can act in one way or another, but having an idea around the odds of success helps make it an informed opinion.

LUCK AND CALCULATED RISK

The real world is messy and risky. Take gambling. Isn't gambling just blind luck? To the amateur, yes. But to those who understand the nuances of games like poker and blackjack, the right game can be gamed. In the semifictional book

Bringing Down the House, a group of MIT students masters the game of blackjack through intricate multi-person, card-counting schemes.[2] The group makes millions before Vegas casinos catch on to their ploy. They didn't break the law—but they exploited known weaknesses in the game to take calculated risks.

We all know people we describe as risky. Perhaps they started a business with their life savings. Maybe they staked their reputation on a controversial political cause or quit their stable job to pursue a passion. Why are these things risky? Because they can all lead to failure.

But failure is not the only option. People take calculated risks because there is a possibility of success that would otherwise be unachievable. The business could take off. Championing the cause may lead to desired legislative changes. The new job may fulfill your dreams.

Think about businesses that analyze risk/reward or cost/benefit of various actions. Why do they do this? Because every situation is not cut and dry. For every action, there are potential risks and potential benefits. When you analyze the situation, you ensure that you understand the risks before moving forward. This doesn't mean that failure won't happen. It just means that you've made a determination that a certain level of failure is acceptable as you attempt to achieve something greater.

LUCK AND PROBABILITY

"Playing the percentages" is a common phrase in the sports arena. In baseball, managers always bring in the southpaw pitcher to face lefty batters late in the game. NFL teams rarely "go for it" on fourth down (though every child who has ever played *Madden* on PlayStation risks it regularly). ESPN even considers Texas Hold'em to be a sport and airs poker tournaments. Many casual watchers don't fully grasp the probabilities being flashed next to a pair of tens, but the players do.

On the screen, these athletes and gamers seem stone-faced. In reality, it takes a steel stomach to play the percentages. Anyone can get lucky for a short period of time, but when we discuss luck as a Fuel, we're not talking about pure happenstance. We're talking about planned luck or even weaponized luck, if you will.

Let's use a simple fictional example. You are invited to a competition on prognostication of future events. Your host suddenly breaks out the equipment for the main event: a single quarter. He says that he is going to flip the coin one hundred times. Your task is to guess the number of heads within two to win double your money. How many people would choose ninety? Zero. Why?

The probability of that event occurring is so low that it's an automatic loss. You would likely choose somewhere close to fifty. In this example, you weaponized luck. You used your knowledge of mathematical probability to game the system and lead to a more likely victory.

Think about scientific studies. If you trace the annals of your memory back to your high school or college statistics class, you may remember something called an alpha value. Alpha is a probability of the experimental results being untrue by chance. When scientists analyze data, alpha is set to a level at which they feel comfortable that the results are very unlikely to happen by chance.

You could say that scientific theories across multiple fields are based on complex principles of luck. Everything you know about the world could be an illusion if these scientists are wrong. That's why replication of findings is so important for scientists. Flipping a coin one hundred times and getting seventy-five heads one time is interesting, but getting that result four times in a row may indicate that you have a special coin.

LUCK AS A RESOURCE

Firestarters who use luck as their Fuel aren't suckers. Sometimes the potential outcome is so likely that pursuing an action based on luck is well justified. Will playing the odds always lead to victory? Of course not. Some choices in games lead to wins, and some lead to losses. But those who don't play the game have zero percent chance of winning, and Firestarters don't like those odds.

Firestarters aren't born lucky. They manufacture it. They're builders. They analyze many characteristics of a situation before acting. When that information includes pros and cons or a probability, they are more informed. They have knowledge of the situation's potential.

SPARK YOUR THINKING

1. Do you consider yourself lucky? In what ways?
2. How does luck affect your decisions in your professional and personal lives?
3. Has your orientation toward risk prevented you from pursuing things that are important to you?
4. How do you calculate risk? Is it different professionally versus personally?

Chapter 14

SOCIAL CONNECTIONS: INSPIRE LOYALTY

To succeed in this world you have to be known to people.
—Sonia Sotomayor, US Supreme Court justice

Mom was right. Who you know *does matter*. Corporations and professionals around the world don't spend countless hours networking because it's a worthless practice. Social media influencers looking to capitalize on their followers don't spend hours cultivating dialog because it's fruitless.

Think about a cause that you're passionate about. Would your friends and family donate to a charity if you asked them to support it? When you are socially connected to people, your passions are important to them. Your passions drive their actions. If your social circle ignores your passions and thoughts, it's not really *social*. Being social is not unidirectional sharing on Facebook. Being social implies a give-and-take and a true concern for the interests of someone else.

THE PERSONAL ROLODEX

Some people collect baseball cards. Others collect business cards (or their digital equivalents). Others collect social media followers. People who look to tap social connections as a resource always try to stockpile them. The more people you have in your immediate network, the more people with the potential to act when you need it. You can probably remember a time when you were in a pinch. You sent out a flailing social media post for help, and someone completely unexpected responded with support. That situation was made possible by collecting connections. You had a stockpile of resources available for exactly that type of situation.

Building social connections requires action. Social media users who share, like, reply to, and comment on others' feeds generally build more loyal and engaged connections. Why? Because they are signifying they actually care about their followers. To build stronger social connections, you must actually be social.

NETWORKING POTENTIAL

When Ryan Seacrest set out to start a reality show about the exploits of a family, he reached out to a casting director and discovered the Kardashians were interested. He came up with the idea to send a cameraman to their house during a regular get-together and found the chemistry he was hoping to see. Members of the Kardashian clan weren't regulars in his social circle, but he was able to link with the past acquaintances through some professional networking. *Keeping Up with the Kardashians* (and its spin-offs) went on to become a hit franchise for the E! network.[1]

The Kardashians' collective business empire has expanded exponentially beyond television, largely because of their ability to cultivate and migrate their following from one platform to another. They recognized that each moment has networking potential. Whether through social media presence, traditional media spotlights, or other outlets, they find a way to bring an audience to the table.

Networking potential is about leveraging a social moment. Each day you meet someone new or spot something in your social media feed that interests you. Do you capitalize? How often do you click with new people on the train or at a business function?

Some people may be uncomfortable in these direct social situations. So for more introverted people, how often do you convert social media followers into your fans through your interaction? In a world with a 24/7 social media presence, networking potential exists for all people all the time. All it takes is the simple action of reaching out.

SOCIAL CONNECTIONS AS A RESOURCE

Many social connections are never tapped because it feels "icky" to *use* someone. Those who aren't natural salespeople often shudder at the thought of asking someone they know to help achieve something. It can make your connections feel like pawns, rather than potential friends or colleagues. If you have this feeling, try flipping the script.

If a close friend or colleague called you and asked for your assistance, would you help? Probably, if you could. Would you feel used? No. Why? Because that's what friends do. You serve as mutual resources for each other. It's the nature of the bond.

This is why we emphasize the social aspect of the social connection. If you always say yes, but the other person in the equation always rebuffs calls for help, you feel used. Why? Because there's a social expectation of reciprocity when you're truly connected with someone.

So the feeling of "ickiness" when the tables are turned should only occur in a one-sided relationship. When you *use* your social connections as resources, ask yourself if you would do the same for them. If you wouldn't, then maybe your icky feeling is justified because you're truly using them. If, however, you would truly reciprocate in similar situations, there is social equity that should eliminate any negative interpretation of the situation.

One of the people featured in our profiles is Patrick Ip. Patrick, who worked for Google and now has a start-up focused on matching brands and influences, is known as the "How Can I Help?" person among his large international network. If you want to find a resource, you ask Patrick. Ballers, a closed Facebook group he started, has almost two thousand people connected throughout the world who are committed to answer the "how can I help" question. But Patrick also added a twist to Ballers when he established dinners throughout the world so members of the group connect both virtually and in person.[2] That is powerful social connecting.

SPARK YOUR THINKING

1. Are you more comfortable establishing virtual or in-person social connections? Why?
2. How often do you ask your connections how you can help them?
3. Do you believe that meeting person-to-person is a necessary component of developing strong social connections?
4. What wins have you had from social connections? Are you able to replicate these successes?
5. What opportunities have you not been able to capitalize upon because of lack of social connections? How would you fix this?

ACCELERANTS: WHAT SPREADS THE FIRE

The Firestarter Framework

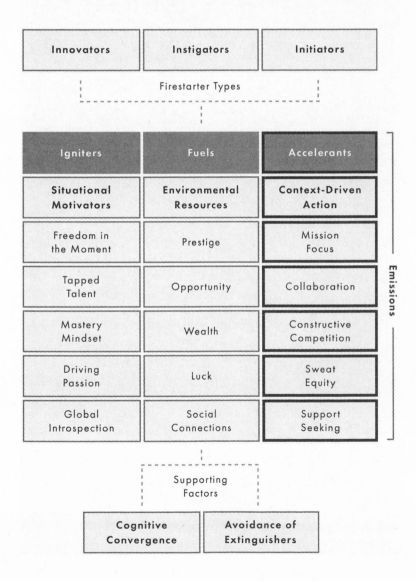

Chapter 15

ACCELERATING YOUR FIRE

Afire doesn't spread itself. It needs some force, something to accelerate it in one direction or another. So do Firestarters. The need for action from a Firestarter never stops.

Yet, Firestarters don't just move for the sake of moving. The movement is purposeful.

Paul Eder, coauthor of this book, and Everett Marshall have spent a great deal of time studying how organizations move. They laid out a model for organization-level strategic movement that has applicability for Firestarters as well. Known as the Socially Balanced Strategy, it draws from social science research on a construct called social value orientation.[1]

What is this? Researchers investigating social value orientation distinguish between actions that one performs for one's own benefit versus actions that one takes to benefit someone else. There are five main orientations that result from combinations of caring about one's own outcomes and caring about others. While there are theoretically more orientations, as a general rule people tend to avoid masochistic action that inflicts harm on the self.[2]

The five orientations include the following:

1. Altruism: maximizing outcomes for others
2. Cooperation: maximizing joint outcomes with others
3. Individualism: maximizing outcomes for the self
4. Competition: focusing on surpassing others
5. Aggression: focusing on disrupting outcomes for others

Eder and Marshall argue that the most effective strategies for organizations embrace all five orientations, leading to what they described as social balance. Note that in a business context, they reframed aggression as an orientation toward disturbing the status quo. Organizations who strive for social balance are more likely to develop a fit between each employee's natural orientation and one of the orientations adopted by the organization.

In addition, social balance occurs in both internal and external contexts. Organizations strategically overhaul internal operations or capitalize on relationships with customers and competitors. Similarly, strategic actions of Firestarters have internal and external targets. They focus on the betterment of the self as a vehicle for change or the betterment of society as a target of the change.

All people have a default social value orientation. Their actions in average situations are likely to be most aligned with their preferred orientation. For example, a cooperator would generally look for opportunities to maximize joint outcomes with someone else. A competitor would seek ways to maximize the ability to claim a "win" over the other person. But people should not always constrain their actions to the social value orientation that is most comfortable to them. They must adapt and apply the action that the situation requires.

The figure below shows how the different social value orientations align with the Accelerants in the Firestarter Framework. We call your attention to two anomalies in the model. First, "support seeking" is an action that aligns with *all* preferred social value orientations. You can seek mentors, guidance, and supportive environments for either ego-driven or pro-social purposes.

The model emphasizes a general "disturbance" orientation that permeates a Firestarter's approach to the world. Accelerants spread your fire by breaking down barriers. Innovators, Instigators, and Initiators all disturb the status quo in different ways.

For Firestarters, the situational context drives action. Sometimes, you look for the right partner. Sometimes, you stare directly in the face of a competitor who must be overcome. Other times, you push forward out of sheer will and effort. The situation chooses your approach. Your job is to be aware and exercise the best action in the moment to give that situation what it needs.

Firestarters act when others don't. When situations call for difficult choices, they make the choices that others refuse to make. When they need other people, they humble themselves and look for the best partner or mentor to drive the mission forward.

Sometimes people can misinterpret Firestarters as ego-driven and obsessive. Are these descriptions accurate? Well they can be, but they don't paint a complete picture. A better word to describe Firestarters is *impact-driven*. They live the change that they wish to enact in the world. Their lives are a constant flow of Igniting, burning Fuels, and Accelerating. The cycle is dizzying to watch, and their pace of action can burn others out.

Accelerants: What Spreads the Fire

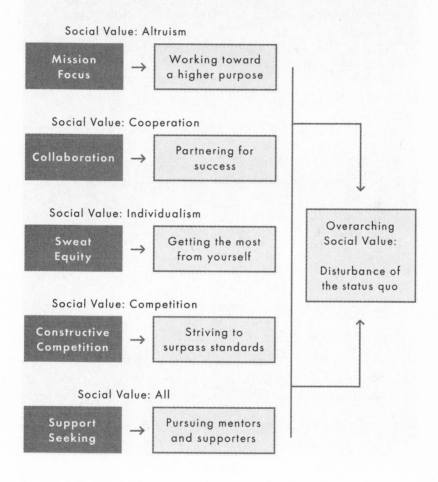

Fig. 15.1. Accelerants.

If you want to be a Firestarter, you have to move. You have to act. If you want to make a difference, you must live like no one else and strive to achieve goals that others view as impossible to reach. Firestarters aren't better people, per se. They're just better at being more effective people.

SPARK YOUR THINKING

1. What is your default social value orientation?
2. Do you agree with the concept that situational context drives action? Why?
3. How has support seeking benefited you? What more can you do in this area?
4. What percentage of your actions are performed for your benefit versus the benefit of others?

Chapter 16

MISSION-FOCUSED BEHAVIOR: TAKE THE LONG ROAD

On December 14, 1911, a group of Norwegian explorers led by Roald Amundsen became the first people to reach the South Pole after months of voyaging and hiking.[1] Amundsen faced stiff competition from Robert Falcon Scott, an English explorer whose plans for the South Pole had been much longer in development.

In the end, Amundsen's team reached the South Pole first and planted a Norwegian flag proudly in the ground. The team overcame many internal tensions and the threats of constant frostbite. After reaching the South Pole, the most difficult days were yet to come—there was no way to announce to the world that the destination had been reached. It took months of brutal travel back to civilization before Amundsen was able to announce his accomplishment to the world. Can you imagine achieving such a great accomplishment only to realize that your real journey has just begun?

MISSION FOCUS AND THE BETTERMENT OF SOCIETY

Having a mission isn't something that's easy. You don't have a mission until the mission consumes you. You don't just wake up one day and decide to have a mission, throw a dart at the board, and skip out the door. You put your pants on one leg at a time day after day and gradually develop a sense of purpose. And that sense of purpose gives rise to the vision of a greater life mission.

For many people, a mission involves helping those less fortunate. Think of Blake Mycoskie, founder of TOMS Shoes. TOMS focuses on helping those in need. Since its inception, TOMS has accelerated beyond shoes to eyewear, drinking water, and safe birthing practices. How has it managed to grow? TOMS has a consistent mission and a simple message—"One for One"—that ensures each purchase helps a person in need with an in-kind product or service.[2]

Clarity of mission is important for acceleration. If you have a mission, but others don't understand or your actions contradict it, then it will be less contagious. TOMS's philanthropy-based business model spreads because it was understandable, and Blake's actions are living proof that it isn't a sales ploy.

One of the areas where mission focus is most recognizable is through acts inspired by faith and spirituality. People with a deep-seated belief in God or a spiritual core often describe the importance of being "called" to a certain course of action. Sometimes, this occurs through a religious affiliation. Other times it is driven by spirituality not directly linked to a denomination. And, of course, there are Firestarters with a mission focus that does not connect with faith and spirituality.

Nonetheless, many of the Firestarters we talked with have deep centers of faith and religious affiliation. They represented many religions and degrees of observance. We did not ask if faith and spirituality were part of their Firestarter makeup, but they repeatedly told stories that emphasized this component of their mission. If you are looking for inspirational stories of faith, we urge you to read the profiles of Innovators Dr. Pernessa Seele and Dr. Barbara Hutchinson in chapter 27. Certainly, many people will come to clarity around missions that aren't faith-based, but it can be helpful to have a foundational mission built on faith.

Faith doesn't always have to be seen in a religious or traditional spiritual sense. Pastor Andy Thompson of the ten-thousand-plus-member World Overcomers Christian Center describes faith as "not just a spiritual, but a practical, concept! It is a life concept, whether it is getting in the car for a quick drive or going to sleep at night and planning what you are going to do the next day; you have to have faith so that you will make it to your next destination and that you will wake up in the morning."[3]

For Firestarters, whether they have faith in a higher power or in the power of their own action, it centers them and allows them to capitalize on their Igniters. Their faith doesn't make them immune to what is going on around them, but it allows them to stay focused on their destination despite the circumstances.

Just think of Dr. Martin Luther King Jr. Instead of confronting police officers during the Selma march, he led the marchers in kneeling, praying, and retreating.[4] Dr. King's faith guided him to believe this was the right course of action, and history proves him right.

MISSION FOCUS AND FLOW

A researcher named Mihaly Csikszentmihalyi has spent years investigating the concept of flow. Flow involves a complete immersion in activities where one focuses intently and seemingly forgets the outside world.[5] Flow combines an intense form of excitement with focus on achieving a result. People in a state of flow often forget about time and just enjoy a state of complete situational mastery.

Flow drives individuals to seek progressively more difficult tasks in their chosen domain. Continued mastery over iteratively challenging events becomes a driver of additional action. A state of flow energizes, complements, and builds upon your mission.[6] Required skill and challenge combine in just the right way to propel action. When you find yourself in flow, you feel actualized in the moment and forget lower-level needs in the face of a larger accomplishment that consumes you.

Csikszentmihalyi also proposes a notion of shared flow where multiple actors come together in a joint consuming experience (like a rock band's jam session).[7] Two people or a group can jointly enter a state that allows them to simultaneously work toward their disparate missions simultaneously and effectively.

MISSION FOCUS IN ACTION

Focus isn't silent. Focus isn't passive. Part of the reason why mission focus accelerates your effectiveness is because it can be communicated. It is message ready. Research shows that a communicated vision has a direct relationship to business growth among entrepreneurs.[8]

A mission that resides solely in your head isn't a mission. A mission that only you live isn't alive. Missions are social; they involve other people. A focus with only one pair of eyes is blurry. It's the act of communicating that makes these things a reality. If others don't know what you're trying to achieve, how can they help you? They can't. Make it easy for them by taking initiative and letting them know.

SPARK YOUR THINKING

1. What is your mission in life?
2. Do you believe you can accomplish it?

3. Can you describe moments when you were in "flow"? How can you replicate those times?
4. What drives your faith in this focus?
5. How are you articulating your mission to others?

Chapter 17

COLLABORATION: INSPIRE EACH OTHER

Alone we can do so little; together we can do so much.
—Helen Keller

James Patterson is a book factory. He has produced more book sales than any author in the twenty-first century, and his influence crosses age ranges, including adult thrillers, middle-grade novels, and picture books. In 2015, he and his team had thirty-six books land on the *New York Times* best-seller list.[1]

How does he do it? He bypasses ego for action. Patterson works with more coauthors than most best-selling authors ever think about. For many projects, he creates book visions and plotlines and hires other authors to bring the vision to fruition (with his support, guidance, and red penning).[2] His stamp of approval is certainly on every work, but he's not the type of author who requires his art to be completely his own.

James Patterson didn't always have an entire team. He had to earn his way to the top through his own writing first. But once he got there, he was able to leverage his advertising background to build a coalition around his brand with his publisher. Not all authors do that. Patterson recognized the importance of collaboration in promoting his goal of reintroducing books to those who have abandoned them.[3]

FINDING YOUR INNER COLLABORATOR

What is collaboration? Collaboration involves both joint effort and joint responsibility. You must understand your role as a collaborator, but you must also accept some degree of ownership for a partner's outcomes.

You're not a good collaborator if you don't care that the other person succeeds. James Patterson doesn't just capitalize on the backs of others. He creates a platform for them that they use as a jumping-off point to become best-selling authors in their own right.

Good collaboration also involves recognition of what you offer to a situation. What are your strengths? The answer may be a little more complicated than it sounds. The same skill set brought to different situations or when combined with skill sets of others may yield vastly different results.

Researcher Robert Ployhart recently introduced the concept of human capital resources. This line of study in organizations focuses beyond individual-level differences to how an individual interacts within an organizational system. You do not perform in a vacuum. Coworkers and other stakeholders have their own skill sets that can be complementary or detrimental. In this conceptualization, an individual is either more or less than the sum of her own capabilities, dependent on what others in the environment bring to the table.[4] Being your best collaborator means understanding the uniqueness of each situation and how you can support others given the constraints or enhancers of the moment.

Collaborators don't steal others' ideas, take advantage of people, or sit back while others accomplish their tasks for them. Collaborators take action to ensure that everyone with whom they work can enjoy the maximum potential outcome.

COLLABORATION AS PARTNERSHIP

The number of partnership businesses in the United States grew from 1.3 million in 1980 to 3.2 million in 2011.[5] What does this mean? As the world becomes more connected, millions of people recognize the importance of collaboration in starting business enterprises. Why? Because we need other people's talents and time to be successful.

It would be great if you were born with every skill set you need to succeed. You weren't. It would be wonderful if other people jumped at opportunities to help you by virtue of your wonderful charm and charisma. They don't (at least for most people).

What do you have left? The ability to be a good partner. You bring talent to the table. You also bring gaps. The gaps could be based on actual skills, or it could be based on time needed to fully exercise those skills (which you don't have). How do you react? Well, Firestarters find partners. They find people whose skill sets and/or availability complement their needs in the situation. They develop an agreement of time and effort that they will jointly expend to achieve the result. Then they work together to carry out the plan.

COLLABORATION IN ACTION

Human beings are social creatures. We like to work with each other in harmonious environments. Collaboration is a naturally adaptive phenomenon. We work together to get things done.

Firestarters recognize the need for collaboration in reaching their ultimate goals. Perhaps you want to start a business but don't know a lot about finance or information technology. Perhaps you have a passion for a charity, but you don't have the social prowess to ensure a thriving fund-raising base. Firestarters overcome these challenges by inspiring others to join the fight. They recognize the talents that others bring to the table and find ways to maximize outcomes for themselves and their partner.

Collaboration is more than tapping the talents of someone else. It involves "living the mission" with someone else. Every person has unique needs and goals. True collaboration includes ensuring that those who partner with you are able to achieve their goals as well.

SPARK YOUR THINKING

1. What would you do tomorrow if you had the right partner to assist you on your journey?
2. What skill sets would complement yours?
3. How would you rate your collaboration skills?
4. What are your most successful ways to get others to collaborate with you?
5. What examples do you have in your professional and personal lives where collaboration broke down? In retrospect, how would you have done it differently?

Chapter 18

CONSTRUCTIVE COMPETITION: RISE TO THE OCCASION

I have been up against tough competition all my life. I wouldn't know how to get along without it.

—Walt Disney

A common elementary school daydream for young boys and girls unfolds like this:

It's the bottom of the ninth inning. Bases are loaded. There are two outs. The count is zero balls and two strikes. You are one swing away from hero or zero status. Your knees are buckling, as anyone's would, even the most stone-faced among us. The pitcher receives the sign and sets himself for the delivery.

Your moment has arrived. Time stands still or at least slows to the speed of one mental video frame at a time. You can see the spin of the ball. You can see its red seams clearly and the off-cream scuffs from prior hits.

All that exists is you and the ball. And the bat, of course. You swing mightily across the plate, and it's a walloping shot that clears not just the fence but also the entire stadium. The crowd goes wild. You did it! You did it! Grand slam!

Now that's an ultimate moment of rising to the occasion. But while many dream about that moment, few actually live it. It's not that people don't bat in the bottom of the ninth (figuratively and literally). Many do. But few gifted people are able to own that quintessential competitive moment. Some even own it repeatedly. Think about Michael Jordan who led the Chicago Bulls to *two* three-peat championship runs for the National Basketball Association in the 1990s. He was able to leverage his natural skills in a competitive environment where many top players struggle to maintain their consistency. Life, however, is made up of many less dramatic "rise to the occasion" moments. How we triumph and how we learn from not winning are crucial in developing and expanding Firestarter potential.

COMPETITION WITH THE SELF

Some people run to beat their own best times. The standard is simple. You ran your last mile in seven minutes. This time you strive for six minutes and fifty-nine seconds. We often think of competition as having both a winner and a loser. That's not always true. Who loses when you're racing against yourself? If the goal is self-betterment, competition with yourself is not a zero-sum game.

Self-competition must focus on improvement to be constructive. Imagine you are playing basketball using a child's three-foot-tall basketball hoop. It may be fun to show off your dunking skills once or twice, but if you had to play for hours, it would get very boring. Why? You've already reached your ceiling of performance. There's nothing to improve.

Competition with yourself requires that you haven't achieved your highest personal standard. Competition is not an effective approach for maintaining a performance level that has reached a natural plateau. Competition requires the inner drive to be better *and* the possibility to make that happen.

Earlier in this book, we discussed global introspection and the importance of self-regulation, which involves identifying standards and self-evaluation of achievement against those standards. Those with a constructive competitive approach those standards from an achievement mindset that others don't have.

Standards aren't just "nice to achieve"; they are "must-be-achieved" guidelines. But a failure to achieve them isn't the end of the world. You reassess. Why wasn't the standard surpassed? What did you learn, and how can you apply that lesson to the next attempt to surpass the standard?

COMPETITION WITH OTHERS

We like to exceed our own marks, but exceeding others provides a different cue of competence, doesn't it? People often gain meaningful information about performance by comparing themselves to others. It signals that you have skills and gifts. In fact, research has shown that in order for rewards to make you feel competent, it is important to know the normative performance level of other people.[1] If you don't think you've surpassed others, a reward does not mean as much to you.

Accordingly, rewards gained for surpassing others are most likely to spur interest and passion.[2] In a meta-analysis of the effects of reward on intrinsic interest, Judy Cameron and her colleagues found that the category of reward most likely to be related to high motivation and free-choice behavior was contingent on achieving a higher level of performance than others.[3] Many other types of rewards tend to have demotivating effects, but there is an inherent quality in surpassing others that provides the highest amount of reward-driven motivation.

As we've discussed, competition's effect on motivation would be strongest when it's perceived as noncontrolling. Think about it. If you are a salesperson, don't you enjoy a friendly competition with coworkers more than an obsessive focus on numbers from management? Probably.

Why? Because the friendly competition has different stakes. When the outcome of a competitive action is your survival in your organization, you're not really competing; winning is the only option. Everything else is failure. In order to be constructive, competition must allow for an honest assessment of action. Too many constraints can bias this assessment. You may be inclined to pass the blame for not reaching management's goals onto other individuals or the environment. When you pass the blame, you don't learn; you forget purposefully.

CONSTRUCTIVE COMPETITION IN ACTION

Competition is an enhancer of both competency and confidence, but only under two conditions:

1. You emerge victorious and surpass standards that have been identified.
2. You do not surpass a standard but incorporate lessons learned from the competition to apply to the future.

If you don't win and you don't learn, competition sucks for you. This highlights a key difference between Firestarters and everyone else. Firestarters seek to win or learn. Either outcome is acceptable. Others seek only not to fail. Cues of competence provided by competition drive the Firestarter to a higher level. You can't pay attention to cues of competence if you always have one eye on failure.

SPARK YOUR THINKING

1. What role does competition play in how you look at life?
2. Are you more competitive with yourself or with others? Why?

3. Have you set certain standards in terms of how you approach what you want to do and how you will achieve it?
4. Do you really believe that losing is okay as long as you learn from it and incorporate lessons into future actions? Why or why not?
5. Are you a failure-focused person? If so, how has it helped or hindered you in achieving your goals?

Chapter 19

SWEAT EQUITY:
EXERT EFFORT TO OVERCOME OBSTACLES

*There may be people who have more talent than you, but there's
no excuse for anyone to work harder than you do.*
—Derek Jeter, former New York Yankee

For nine seasons, Jack Bauer thrilled his targeted eighteen-to-forty-nine-year-old audience on the television show *24*. Each episode represented a portion of the same day culminating in the completion of the twenty-four-hour cycle in the final episode. To the audience, Jack was a kind of relatable superhero, literally running for twenty-four hours straight. Jack was smart and savvy, certainly. But where he shined was often in outlasting the opposition more than outthinking them. Pure, raw effort made Jack stand out from the pack and made him a go-to resource even when others questioned his methods or gruff persona.

For Firestarters, life is a series of twenty-four-hour sprints. Every day invites effort, and Firestarters give more effort more consistently than others. Their sweat is harder earned and more impactful by design. A Firestarter's effort is strategic; they don't waste action.

EFFORT AS PERSONAL REWARD

There is a difference in how people perceive effort. Is the feeling of exertion viewed as a necessary evil or an internal joy? How much do you crave effort for effort's sake? In 1976, Martin Seligman laid out evidence in support of a theory of learned helplessness.[1] Simply put, through a series of aversive experiences, people and animals can learn over time that their actions don't influence outcomes in a given situation. They then generalize this lack of outcome efficacy to other situations where their effort (or lack thereof) really could make a difference. They don't act even when action matters.

Earlier, in chapter 2, we discussed the tenets of learned industriousness, a research framework laid out by Robert Eisenberger.[2] From a learned industriousness perspective, people who experience positive outcomes as a result of great exertion come to view the exertion itself as a rewarding experience.

Eisenberger and his fellow researchers demonstrated that people can learn that increased mental effort is desired and supported. For example, Eisenberger's research group found that fifth- and sixth-grade students increased creativity in one task (picture drawing) when novelty had been previously encouraged in another task (novel uses for physical objects). This association occurred when expectation of creativity was made salient by promise of reward.[3]

In school, children are often rewarded for sitting down, being quiet, and being good listeners. This pattern of reward would suggest that conventional, rote, rule-following effort is what is expected. Effort is something that can be learned, but even this learning is nuanced. The specific type of effort that you exert (conventional vs. creative), and therefore find enjoyable, also depends on how you interpret the situation and how others have shaped your expectations over time.

Firestarters want to break a conventional mode of thinking that has been tossed around their necks by society. Earlier, we discussed the importance of self-reward. That is important. But it is equally important that the expectations of yourself involve high effort and willingness to break the conventional mode. High effort by itself is great. But high effort in pursuit of unconventional ends is what makes you stand out.

SWEAT EQUITY AND RESILIENCE IN PURSUIT OF MISSION

Imagine these scenarios. You are in a paper-stacking competition. Exciting, right? You must stack as many sheets as you can in five minutes. In scenario one, you stack one hundred, seventy-five, and fifty papers in successive rounds. In scenario two, you progressively improve your stacking in each round. First you stack fifty, then seventy-five, then one hundred.

In both scenarios, you stack the same number of papers, but scenario two would be more motivating. Why? Studies show that individuals receiving information that provides evidence for progressive mastery of tasks are more likely to be efficient and perform at higher levels than those informed of progressive decline.[4]

Your reaction to the sweat equity that you've built up is dependent on the results you produce. Additionally, the psychological results of your effort are a feedback mechanism. The same overall effort in the two scenarios yields different levels of motivation.

Effort that provides evidence of improvement inspires further action. This is why resilience matters. People who suffer through negative events are in danger of losing hope. But evidence of improvement in the situation helps to heal the psyche. You feel that your effort is driving results. When you mentally travel from a low point to a high point, the ensuing inspiration eclipses the negativity of the initial downfall.

Contrast those who focus on progressive improvement with those who give up after a failure. Consistent avoidance tactics do not allow you to recognize when reality changes. You won't be able to notice the improvement in results because you never acted to improve them. Occasional failures that you overcome effectively inform you that your effort will pay off. You come to see effort as worthwhile when overcoming similar obstacles.[5]

SWEAT EQUITY IN ACTION

We all know the feeling of sweat, the mastery of a passion through endless hours of toil. We all know the feeling of fingers bleeding (some literally) and knuckles dragging from exhaustion. We all know what effort feels like. Some of us like it. Why?

Human beings aren't masochists. We choose direction based on perceived positive outcome for us and those close to us. Effort will not be your choice unless you see some value resulting from that effort. Firestarters who recognize a personal connection between effort and outcome view that effort as instrumental to success, a pain to be endured for greater good rather than just a pain.

Researchers have defined three types of personal initiative that align with the sweat equity:[6]

- Self-starting: doing things without being told
- Proactivity: planning ahead for problems and opportunities
- Persistence: overcoming barriers that arise

Firestarters put all three types of initiative into action. They get moving, think ahead about where to act, and persist in the face of failure. Some people cringe at the idea of self-starting. Others fall so far behind that they can't even contemplate being proactive. And still others give up when the situation looks bleak. Firestarters don't let physical and mental blips impede their action. They know that the next round of any effort is the one that could propel them over the top.

SPARK YOUR THINKING

1. When was the last time you gave up after a failure? Why?
2. What would the results have been if you hadn't given up?
3. How would you handle the scenario differently today?
4. In what area do you feel you need more focus—self-starting, proactivity, or persistence? What can you do to improve?
5. Do you get more reward from progressive mastery of tasks? In what ways?
6. What are the best examples in your life where positive outcomes made you work even harder?

Chapter 20

SUPPORT SEEKING: HUMBLE YOURSELF

You can find a mentor; you have to ask questions, you have to show interest in what the other person is doing. You have to have curiosity—I think that people appreciate that and will want to help you.

—Nina Garcia, judge on *Project Runway*

On the television show *The Voice*, singers from around the country compete for slots on teams to be coached by some of the most talented artists in the music industry like Adam Levine, Alicia Keys, and Blake Shelton. Why do they do this? Because they want to be famous? Sure. There's certainly a bit of ego involved. However, if you listen to the contestants during on-camera interviews, one of the most pervasive themes is that they want to be *better*.

Working with great artists is a wonderful ego trip, but the biggest benefit is the ability to learn something from people whose success the contestants wish to replicate. The show uses the term "coach" purposefully. The artists have real access to the thoughts, opinions, and experiences of industry experts. Those who succeed humble themselves and realize the importance of learning from the coaches to improve their odds of success.

SEEKING MENTORS AND TEACHERS

Many of the Firestarters we interviewed discussed the important role that teachers and mentors have played in their lives. What we found most intriguing was how these Firestarters actively sought teachers and mentors from multiple places around them. They wanted to learn and didn't have a need to be seen as the smartest person in the room.

When you meet people who work in a different field than you, do you ask them questions about their work? Do you truly want to understand what people with other jobs do? What accomplishments do they seek? What are some types of people they come across? Firestarters want to know these things. Not just for knowledge's sake, though that is important. They want to find out how people who succeed in diverse fields achieve that success. By seeking information and advice across spectrums, they are able to develop and enhance an internal worldview that paints a picture of the broader secrets of life.

Students with supportive, warm teachers have accelerated levels of motivation.[1] This isn't surprising. Think about college when students have a choice of instructors. Students seek those whose styles are more supportive of their learning experience. This isn't magic. It's forethought. Think about what you want to achieve. Identify those who can help and seek their advice and guidance.

SEEKING SUPPORT FOR OUTCOMES

Many avenues of management research highlight the role of social exchange. The individual and the organization have a symbiotic relationship. When the organization does something good for the individual, the individual in turn feels obligated to do something good for the organization. Research findings from a variety of industries support this basic premise: organizational support yields results.[2] Good treatment yields good performance.

Not only do people prefer supportive environments; they seek them. Organizations across the country focus on the idea of employee engagement, which represents an amalgamation of many related social exchange concepts. In sum, the availability of supportive supervisors, effective practices, and growth opportunities result in employees who contribute in positive ways. Why? People want to work in an environment where they feel supported.

What does this mean for Firestarters? The road to your highest achievements leads through environments that support you. Some people are reluctant to leave bad situations. Some people burn themselves out for bosses who don't care. Those situations are draining. Firestarters pay attention to the environment. They identify situations conducive to their growth and supportive of their missions, and they pounce. They know that their success is maximized in environments that value their contribution.

SUPPORT SEEKING IN ACTION

Being mentored is not a passive activity. Often organizations will set up programs where new employees are paired with experienced employees. Mentoring programs allow those new to the culture to ask questions and learn. This type of mentoring is great for the new employee and likely leads to high-quality onboarding experiences. However, in that environment, you experience a program created for you by someone else.

For Firestarters, life is a series of ongoing mentoring and support. Here's the rub: you must constantly seek your own mentors. Sometimes it will be easy like when you are randomly assigned to a class with a talented teacher. The teacher is a captive mentor who is paid to answer your questions. So take advantage. Other times, access is more difficult—such as when you read a book by an expert (or perhaps three experts) whom you respect. How do you get in touch with the person without seeming stalker-ish? You can reach out via social media if the person is highly responsive. You could also research speeches or lectures that the person gives or conferences he or she plans to attend. Then show up. Be an active audience member, or seek the person's counsel for some active questioning if you both attend the same conference events.

As one Firestarter said during an interview, every person you meet from the janitor to the president has something that they can teach you.[3] The key is positioning yourself to ask the right questions and then asking them. To use the input of mentors most effectively, you must truly care how the other person's worldview integrates with your own approach to life.

Seeking support also means recognizing unsupportive environments and changing them. If the people in your organization drag you down, what do you do about it? When Firestarters assess the value of an environment they determine what they can learn and how they will be supported. If there's no one to learn from, they seek mentors elsewhere. If people in the situation don't value their contribution, they scan the environment for an opportunity better suited for the impact they desire to make.

SPARK YOUR THINKING

1. Do you have mentors who care about your journey?
2. Do you feel an obligation to continue helping others even when they show they don't care about you? If so, why do you allow that?
3. Why don't you move or change the situation?

EXTINGUISHERS:
WHAT THREATENS THE FIRE

The Firestarter Framework

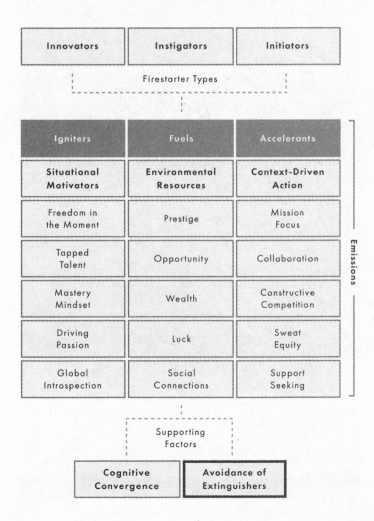

Chapter 21

AVOIDING AND OVERCOMING EXTINGUISHERS

I t's a windy day, and you're celebrating a birthday party. There are several candles on the cake in need of a flame. However, you're outside. The breeze picks up, and it starts to rain. You only have one match. What do you do?

You probably have several potential answers running through your head. *Go inside. Ask someone if she has a lighter. Light the match and shield it with my hand.* Very few of you probably thought, *It's hopeless. Give up and let the elements win.* Why didn't you think like that? Because our minds are oriented toward solutions, not acquiescing in the face of challenges. When there are threats to a fire, we think of ways to avoid them or overcome them. We don't just quickly give up.

In this chapter, we discuss Extinguishers, those elements that threaten to snuff the flames that Firestarters have lit. The figure below illustrates Extinguishers that can threaten the person, environment, and situation. Discouragers threaten Igniters. Limits threaten Fuels. Self-mismanagement, punishers, and ineptitude threaten the situation.

Extinguishers in the person are psychologically based. These are things that freeze your action and prevent movement in the desired direction. Extinguishers in the environments pertain to the nonrenewable nature of Fuels or Fuel limits. Similar to natural resources, the Fuels utilized by Firestarters can run out. Extinguishers in the situation such as self-mismanagement, punishers, and ineptitude represent factors that may impact one person in one situation but will not necessarily translate to other contexts. Something about the particular situation threatens to drag you down.

We approach Extinguishers from a hopeful perspective. When Firestarters are pursuing positive courses of action, they think of Extinguishers as dangers that can be avoided or overcome. The key is recognizing them and acting to thwart their progress before they snuff out your flame.

Extinguishers:
What Threatens the Fire

Discouragers: Extinguishers in the Person	Fuel Limits: Extinguishers in the Environment
Igniters Threatened	Fuels Threatened
Freedom	Prestige
Tapped Talent	Opportunity
Mastery Mindset	Wealth
Driving Passion	Luck
Global Introspection	Social Connections

Extinguishers in the Situation

Self Mismanagement	Punishers	Ineptitude

Fig. 21.1. Extinguishers.

Real fires are extinguished for a number of reasons. You can imagine forgetting to add another log to the fireplace, the dwindling of fuel in an oil lamp, someone taking direct action to snuff a flame, or the fire's inability to spread beyond a nonflammable barrier. But the dwindling of a Firestarter's flame is not inevitable. Fires can be sustained for decades with the correct, continuous mixture of Igniters and Fuels.

Firestarters face, succumb to, and overcome Extinguishers in their lives like everyone else. One key difference, however, is the active planning that Firestarters undertake to avoid the impact of Extinguishers. Firestarters expect Extinguishers. They know they will be discouraged. They know their Fuels have limits. They know they will face uncertain, punishing situations. They know they will make errors.

In tenth-grade biology class, students learn that organisms that face threats have three options: move, adapt, or die. Firestarters explore possibilities for moving and adapting in the face of Extinguishers. They are motivated not to let their fires die.

Every Firestarter we profile in this book faced different Extinguishers. Among the most interesting and stirring stories that we urge you to read are those of Keith Nolan, David Egan, Noah Galloway, Scott Petinga, Karen Benjamin, and Joe Morone. You can find these stories in section 6 beginning with chapter 27.

SPARK YOUR THINKING

1. What are some of the factors that have extinguished you?
2. Have you been more adversely affected by environmentally or psychologically driven Extinguishers?
3. Do you put plans in place to minimize the impact of Extinguishers? What do you find most effective?
4. What motivates you when Extinguishers start to overwhelm you?

Chapter 22

DISCOURAGERS: DON'T LIMIT YOURSELF

[T]he only thing we have to fear is . . . fear itself.
—Franklin Delano Roosevelt

You're afraid. You have no choice. You're not good enough. You don't know what to do. You don't feel like it. Any of these statements sound familiar? Every time you want to move but don't, there's a reason. It's in your mind, and it's blocking you. Your courage is shaken. You remember past failures. You fear how others will react.

Discouragers are aspects of the environment that undermine your *how*. You know what you have to do. You just can't. Discouragers have a particularly devastating influence on the Igniters in the Firestarter Framework. They block you from putting your Igniters into action.

DISCOURAGING FREEDOM

At least once in his or her lifetime, every parent breaks out the phrase, "If all of your friends were jumping off a bridge, would you?" This phrase is meant to discourage choice. Just because you can do what your friends do, it doesn't mean you *should*. The *should* placed on actions in a given situation is a limitation. When there is only one logical or allowable choice (e.g., not jumping off a bridge), you lose some level of freedom.

A lack of control combined with high demands is an exceptionally challenging situation. Researchers examining the demand-control model of work have found that high job demands, when in tandem with a low level of control, yield higher levels of fatigue and strain.[1] Overly controlling environments (even nonwork environments) stifle autonomy and degrade well-being.[2]

The need for security is the enemy of freedom. Think about it. If you hate your job but need the income to support your family, you will be less likely to seek the perceived freedom of entrepreneurship. Why? Because it's too risky. Other people depend on you.

The benefit of freedom in any situation is always weighed against the benefit of increased security. Think of the cyber security concerns faced by government and private sector entities. Employees routinely give up their online freedom to protect their organization's infrastructure and reputation. Your organization may block certain websites that have dangerous coding or messages. You may have to spend an extra ten minutes logging in to your computer with passwords (or multiple passwords). So anything that constrains freedom has a tempering effect. The availability of freedom in any situation is inversely proportional to the need for security. Security often wins out in that equation.

DISCOURAGING TALENT

Some people worry that they aren't talented or that the talents they have aren't good enough to propel success. Think about the most incompetent goofball in your life. Now think about something that person excels at doing (maybe even mooching!).

Now bring the focus back to yourself. If you could find something competent in the least skilled person you know, you should also be able to do the same for yourself. You have talents. The first step is recognizing them.

Often people hide a talent out of fear of what their friends or other observers may think. Think of artists who don't want people to see their work or people who write novels in a diary that they are afraid to have the world see. A purposefully hidden talent isn't useful for propelling you along your path. In order for talent to matter, it has to be recognized. When your talents are ignored, or worse rejected, your momentum is drowned.

When Stephen King wrote *Carrie*, he received repeated rejections from publishers. It is claimed he threw the manuscript into the garbage out of frustration. But his wife fished it out and encouraged him to try again.[3] The book went on to sell millions of copies and launch a decades-long career.

DISCOURAGING PASSION

We have discussed cognitive evaluation theory, where individuals' intrinsic motivation is driven by their assessment of their level of autonomy and competency.[4] Accordingly, highly controlling and negatively evaluative environments have the potential to stifle passion. Indeed, research shows that threats, surveillance, negative evaluation, and deadlines all undermine motivation.[5]

Threats

Threats are controlling. You may be pressured to achieve a certain level of action or outcome with an "or else" contingency. A focus on the "or else" diverts your attention from the true task. Your focus becomes the external constraint rather than the internal drive.

Surveillance

Surveillance is an interesting phenomenon. In the famed Hawthorne studies from the early twentieth century, researchers at a manufacturing organization examined how different aspects of the work environment (e.g., lighting brightness) affected productivity. When results were reexamined, it was determined that any improved performance due to environmental manipulations was more a result of the participants believing their performance had been observed than the actual changes in lighting.[6] Surveillance led to performance in an industrial environment—true. But the results only held insofar as the surveillance was continual.

Surveillance of creative and interesting tasks may be undermining in a different way. Often society rewards conventional behavior.[7] Think back to elementary school where you were encouraged to sit down, shut up, and avoid making waves. Many teachers (but not all) reward students for behaving in a prim and proper way. But creativity, by its nature, is mischievous. It messes with the status quo.

Creativity is suppressed when compliance is demanded. Creativity must be specifically defined as a need to trigger people to adopt behaviors that support it. If a teacher wants to see a creative drawing, it is beneficial for her to ask students to "draw something creative" rather than simply "draw something." By the teacher adding the word "creative" to the demand, the student then knows that outside-the-box thinking is preferred. Otherwise, a student may just draw a stick figure or a house or something else that conforms to convention.[8] When under surveillance, you act in ways that you believe the person watching wants you to act.

For those naturally inclined toward riskier or creative behavior, surveillance in a traditional environment would deter their inclinations. Surveillance drives normative compliance versus outside-the-box action, and thus would inhibit even the best-intentioned Firestarter.

Evaluation

Studies show that positive feedback is helpful for encouraging action and motivation.[9] However, negative feedback is a killer. It stifles motivation. In a meta-analysis, Judy Cameron and her colleagues found that receiving a level of reward that signifies failure at a task results in decreases in time spent on activities.[10] Failure demotivates action. It signifies incompetence, and it defuses passion.

Deadlines

Deadlines are also a constraint that emphasizes compliance over motivated persistence. In 2014, the highly anticipated website for the insurance exchange brought about by the Affordable Care Act was unveiled. The website was slow, and all of the functions were not operational. Using the newly designed healthcare exchange was a frustrating experience for many customers. It wasn't ready for primetime and allowed Republicans to spike the political football.[11] Why did the government team working on the project release such a flawed product? Many people believe it was because of a politically promised deadline. The public expected the website to be released on a certain date, and it would have been politically unpalatable for the date to be delayed. The team's passion for producing the best product may have been undermined by the desire to meet a deadline.

DISCOURAGING MASTERY

Lack of a mastery mindset is a disease that devours from within. If you feel like a master of nothing, you're doing a disservice to yourself. You always have something to feel confident about. Find it and embrace it.

This is difficult when you have a bad day or a series of bad experiences. When people experience negative events, they look to other events that precede or predict the aversive outcomes. Accordingly, a certain degree of trepidation becomes associated with all of the conditions that may predict negative outcomes.[12]

There is an old research fable (fictional but still explanatory) about a group of five monkeys in a cage with a banana hanging in the center.[13] The banana is dangling just within a monkey's reach from the top of a perfectly placed ladder. Inevitably, among the group of monkeys, one catches sight of the banana and starts to climb. As he does this, the rest of the monkeys are sprayed with ice-cold water by a researcher. When another monkey attempts the same maneuver, the group is sprayed again.

One of the moneys is then replaced. The new monkey spots the banana and makes his move. The others attack him to prevent another spray. The same process is repeated until none of the original monkeys who was sprayed remains in the cage. Nonetheless, anytime a monkey makes a move for the ladder, the others attack him. No one remains who remembers why they attack, but they do it anyway. They developed a culture of banana avoidance.

The real research story is less dramatic. It didn't involve spraying water or monkeys beating each other. Rather, it involved one-on-one monkey interactions. One monkey inadvertently trained another monkey to avoid manipulation of objects that were more attractive when the monkey hadn't been paired with a monkey who had an avoidance orientation.[14] But over time, multiple versions of the above story have been reported as true. Why? Well, first it is entertaining. And second, there is a grain of truth. Think about it—how much of what you consider as impossible is based on hearsay or the emotional reactions of others? Mastery is not possible if you don't allow yourself the chance to act.

Additionally, holding oneself to dysfunctional, unachievable standards damages the mastery mindset. It's great to want to achieve the best. But the best should not be inflated by overconfidence in your abilities. When you reach to unachievable heights, you effectively never have the feeling of mastering something.

DISCOURAGING INTROSPECTION

Earlier we discussed the concept of role identity, whereby people can maintain several identities within their psyche, but their actions are usually consistent with the identity most salient at a given point in time. Identity has its roots in the concept of self-regulation. When you adopt an Innovator, Instigator, or Initiator identity, you regulate action in accordance with that adopted frame.[15]

However, reality is not always clean. When you're in an Innovator mood, the situation may call for an Initiator solution. The situation calls for something different than what you're willing to give in the moment.

You feel torn, and your effectiveness is drained. A Firestarter doesn't want to feel this way. No one wants to feel this way. Do you? In the end, one of the identities must win. It plays out like an ultimate *Game of Thrones* battle within your mind. If the situationally needed identity wins, you're on your way to overcoming the crisis and getting things done. However, if the couch potato identity wins, you may be laying the foundation for other distractions to snuff your flame.

Introspection also involves outlining the right goals for yourself that match a situation. We discussed the importance of four-dimensional goals that tap your history, your expectations, others' expectations, and normative comparisons to others. When you evaluate your performance but exclude key contextual dimensions, your assessment of goal achievement will be skewed. Focusing on one dimension may drive you to be overly judgmental of your successful or failed results. When you measure your success in judgmental versus informative ways, you do not allow yourself to grow.

SPARK YOUR THINKING

1. How can you become more free so that you are able to ignite your life?
2. Do you work in an environment that discourages your passion? If so, how will you change the environment or your relationship to it?
3. How do you react to negative feedback? In what ways can you counteract the effect it has on you?
4. Have you set unrealistic standards that threaten your mastery mindset? In what ways, and how can you fix it?

Chapter 23

FUEL LIMITS: DON'T RUN OUT OF GAS

Rule No. 1: Never lose money. Rule No. 2: Never forget rule No. 1.
—Warren Buffett

I've learned. . . . That opportunities are never lost; someone will take the ones you miss.

—Andy Rooney

We're not that much smarter than we used to be, even though we have much more information—and that means the real skill now is learning how to pick out the useful information from all this noise.

—Nate Silver

uels run out. This is true of fossil fuels. This is true of batteries. This is true for Firestarters. This axiom is also the reason why Firestarters never stop their search for Fuels. They aren't content to rest on confidence in a momentary stash. They know Fuel is finite. They must find more or face the risk of letting their fire burn out.

Accordingly, each of the Fuels we discussed in section 4 has limits that are important to recognize. When you understand the transient nature of Fuels, you are better prepared to overcome the limits when you face them head on.

LIMITS OF PRESTIGE

In social situations, power only matters to the extent that someone realizes you have it. Think of a celebrity who walks into a restaurant and is provided with a free meal. What happened? She was recognized. The manager inflated her prestige on sight and treated her in a special way.

You control the prestige you wield and the perception of prestige you bestow upon others. A celebrity could choose to dress in a disguise when in public. Accordingly, any special treatment she may normally receive is snuffed. When meeting a celebrity, you could also choose from multiple courses of action: relentlessly fawn over her and ask for a picture or ignore her. By ignoring her, you reduce the power of her prestige.

King Louis XIV of France famously said, "L'état, c'est moi!" It's translated as "I am the state!" He became the textbook definition of an absolute monarch. Many rulers and dictators throughout history have adopted this philosophy of self-inflation. Many also faced subsequent popular uprisings, invasions, or military actions that led to their overthrow.

Think of the CEO of a major corporation. Her power is bounded. Within her organization, her word may be gospel. But transport her to another organization, and her email directives would be worthless. Why? Because her power is tied to a position in a specific organization. You can't give orders to a group that doesn't recognize you as having legitimate or *earned* authority.

LIMITS OF OPPORTUNITY

Opportunities are everywhere all the time. It is the conundrum we all face. Opportunity as a Fuel has limits precisely because the options are so limitless. You can't move in every direction at once and capitalize on all potential successes. You must choose.

The limits of opportunity are contained in the choices you make. In some ways, you must act like a perpetual soothsayer. For every given moment, you read the tea leaves and choose the direction most likely to drive success, as you interpret it.

Facing this reality, you will be wrong. You will be wrong often. But the power of opportunity also lies in its limitlessness. Being wrong often doesn't matter so long as you are able to be right one time in a big way. One victory could overcome years of defeat. Of course, a series of right small decisions can also give you victory. Either way, you cannot allow yourself to retreat. You have to continue to search. Otherwise, the golden opportunity will pass you by, and you will never know what you missed.

LIMITS OF WEALTH

"Money doesn't buy happiness." We always hear people say this. At its heart, this quote points out the separation between the physical world and the mental world. It's not the wealth itself that matters; it's how you react to it and use it.

Wealth is a finite resource. Your bank account is only so large. The amount of money that friends, relatives, and banks are willing to invest in you has limits. For everyone, there is a point at which available wealth is completely tapped.

Even Elon Musk faced this reality. In 2010, he invested his entire liquidity ($200 million) into his business. As a result, he relied on the kindness of friends who supported his personal expenses to the tune of $200,000 a month.[1] Most of us don't have friends who would lend us money to support our need for a private jet, but the principle is the same. Two hundred thousand dollars is a limit for Musk in the same way that a few thousand would be a limit for others. This is why organizations focus so much attention on budgeting and prioritizing. You can't afford to do everything. So you must prioritize to do the things you can with the capital available to you.

When your funds run out, your movement is stifled. Your focus shifts to survival rather than growth. You begin asking, "How can I keep the lights on?" The lights don't drive your passion, but without them your time is up.

LIMITS OF LUCK

Throughout his career, Nate Silver has focused on the power of statistical probabilities to predict performance in the arenas of sports and politics via his website, fivethirtyeight.com. For Silver and others, the 2016 presidential election was difficult to call. Donald Trump's vote margin surpassed prediction models based on polls and violated political norms espoused by pundits and commentators.

Silver's claim to fame and associated reputation is based on the results of his objective, analytical models. For example, using available polling and other data, he perfectly predicted the results of the 2012 election in all fifty states. Fast-forward to 2016. Throughout the primary process Silver published numerous opinion pieces discounting the rise of Trump, even equating a Trump nomination by Republicans to having similar support as the idea that the original moon landing was faked.[2] The same piece was written using a title that suggested polling data doesn't matter. Yes, someone whose livelihood depends in part on trusting poll data wrote opinion pieces suggesting the data was too noisy to be predictive.

This is the double-edged sword for statistical prognosticators. Sometimes the data will be noisy. But data doesn't predict anything by itself. It requires the interpretations, convictions, and opinions of those who interpret it. Across a range of thought pieces throughout the election cycle, Silver made it unequivocally clear that he disagreed with Trump's style and thought Republicans were making a mistake.[3] He ventured into the realm of pundit, instead of pure statistical prognosticator, and he tested the limits of luck. Silver admitted as much in a post entitled, "How I Acted like a Pundit and Screwed Up on Donald Trump."[4]

Probabilities don't matter if you interpret the data to support your preconceived biases. This reality rattled the political prognostication industry, as there were very few models outside of the one put forward by Professor Allan Lichtman that predicted a Trump victory.[5]

Luck has limits, with the most pervasive being that people are flawed and subject to misinterpreting probabilities in any given situation because they refuse to believe that improbable events can and do happen. Think of the Texas Hold'em player dealt two kings. He may bet it all without knowing that his opponent has two aces. When your own hand is good, you discount the possibility that you could still lose.

LIMITS OF SOCIAL CONNECTIONS

Friends help friends. This is a truism. Friends help friends repeatedly *until* they feel used and battered. This is also a truism.

When we described social connections in chapter 14, we emphasized the *social* aspect of the relationship. Your motives must be pro-social, with a focus on benefits for the other person as well, in order for the connection to remain tap-able. Accordingly, asking for a favor or requesting a moment of attention is respectable, but overwhelming someone with requests wears the relationship thin.

With social connections, you must nourish them and never forget that the *give* is often as important as the *take*. The norm of reciprocity holds that when someone does something good for you, you must also do something good for him or her. When this norm is violated, people feel offended.

OVERCOMING FUEL LIMITS

Dwindling Fuel supplies can cause panic. You may get anxious about where your next paycheck is coming from. You may worry that people who were once your friends no longer return your calls. You may feel as if luck is no longer on your side.

Anxiety and worry are strong feelings, but Fuels aren't feeling-based. They are resources. When you run low, the objective is to find more, not to lament those you have already used. Is running low on Fuel fun? No. But fun doesn't matter when your fire is in danger.

If you drive on the highway and the gaslight on your dashboard illuminates, you strategize, right? You think about what you'll do at the next exit where a gas station may be. What if you acted the same rational way when your Fuels are in danger?

If you miss an opportunity, how do you make sure you don't overlook the next? If your prestige wears thin in one situation, how do you rebuild it for your next interaction? If you run low on cash, how do you get more? Every time your Fuels run low, you can choose to panic or choose to act strategically. Firestarters choose action.

SPARK YOUR THINKING

1. What are the Fuels that you regularly tap in your life? What are their limits?
2. How do you know if they are running low?
3. What can you do to supplement the Fuels you currently have available?

Chapter 24

SELF-MISMANAGEMENT: DON'T BE MISGUIDED

It's a proprietary strategy. I can't go into it in great detail.
—Bernie Madoff, financial manager imprisoned for
operating a multibillion-dollar Ponzi scheme

In 2016, more young adults ages eighteen to thirty-four lived with their parents than in any other living situation (32 percent).[1] Why? The reasons vary. Some cannot find jobs either because of the economic situation or unmarketable skill sets. Some have huge debt burden from school loans. Some don't know what they want to do with their lives. Some are at home to help aging parents. And still others have squandered their opportunities like the biblical Prodigal Son, who wasted his inheritance on alcohol and debauchery.

Managing yourself is more than managing your emotions and impulses; it's also managing your goals, your progress, and the Fuels you must use along the way. Accordingly, self-mismanagement is a threat that can strike at any point during your journey. Every situation has an opportunity for mismanagement.

One classic example is the sad downfall of lottery jackpot winners. Often, we hear about lottery winners who blow their chances of financial security and lose it all.[2] The windfall comes, but they don't know how to manage it, and years later they are broke. Like lottery winners, Firestarters must be adept at recognizing the potential for mismanagement in any situation and injecting self-regulatory balance when the threats emerge.

MISMANAGEMENT AND UNSUCCESSFUL SELF-REGULATION

In early 2017, Kathy Griffin was one of America's most successful comedians. She was a regular on CNN's annual New Year's Eve celebration and was considered a comedic force, willing to push boundaries and provide humor through often tongue-in-cheek political and social commentary. Griffin, while often racy, had managed to toe the line, agitating those who didn't agree with her but building an increasingly larger fan base.

Griffin participated in a photo shoot that included pictures of her holding a bloodied fake head of President Donald Trump.[3] The backlash was fast and decisive from foes and fans. TMZ released a story that Trump's youngest son had seen the image on TV and thought something had happened to his father.[4] As a result CNN fired Griffin, corporate sponsors terminated agreements with her, and many venues canceled her shows.

Griffin was aggravated by what she deemed harsh political language and positions by Trump. However her extreme response negatively impacted her brand, financial well-being, and threatened to permanently turn many of her fans against her. Griffin mismanaged the situation and failed to regulate her edginess to a more politically acceptable level. She failed to step back and reappraise her actions. She failed at self-regulation.

So why do people fail at self-regulation? When discussing global introspection, we described research about reappraisers who actively assess themselves and their environment to find the best way to move forward. They learn what they can do to change, integrate it with their self-concept, and they act.

The same researchers also discussed another less successful type of self-regulation: suppression.[5] Suppressors ignore their internal states and focus only on modifying behavior. Suppressors feel like impostors in their own skin, and they mislead others. Their focus is wrong. They change what they do, not what they believe. Suppressors wear an emotional mask and do not let others experience their internal truth. The feeling of inauthenticity that they experience is debilitating. It surfaces through self-doubt, and they become obsessed with negative events. They have difficulty taking bold, firm action and forming close relationships.

Suppressors use self-regulation as a tool to stifle negative truths, rather than to change them. As a result, hard truths aren't addressed. They are ignored, and they lurk in the back of the mind with the constant potential of overwhelming one's emotions and actions.

Think about what you did yesterday. Did you do everything you wanted? Was every action something you chose? You probably have a hard time with this question. Why? Because many of your actions are on autopilot. You act without thinking. You also likely chose not to act in some ways that you would have preferred either because there were other priorities or you didn't think your action would achieve intended results.

You're not a fortuneteller though. Sometimes you fail to act even though you have the ability to achieve what you want.[6] But you didn't know your actions would have been effective. You may have reflected on the past and determined that the most productive course of action was too difficult, or you may have felt that outcomes were unattainable. Either way, you suppressed action or made choices that were not in your ultimate best interest.

How do you overcome this? It sounds cliché, but you must know yourself. Firestarters understand their personal patterns and tendencies more than most. They reflect. Then they plan their regulatory systems around their own human nature.

MISMANAGEMENT OF FUELS

Fuels are finite. You can't continue to tap the same resource repeatedly and expect your outcomes to be stable. Not all squandering of Fuels is due to pure resource limitations. Sometimes, you waste them. You ask friends for favors that don't matter. You blow money on a luxury car. You bet all your chips on black at the casino.

In April 2016, a diverse group of twenty thousand concertgoers embarked on a journey to the Bahamas for a luxury getaway experience known as Fyre Festival.[7] The experience turned out to be a complete flop. Attendees shelled out thousands of dollars, but the experience was far from luxury. The musicians canceled. The lodging consisted of rows of "refugee" tents. Organizers could not afford the top-quality catering they had promised.

Money came in, but it was squandered. Reports circulated that organizers had spent a good portion of funds on promotion, paying as much as $250,000 for individual celebrity endorsements of the event on social media.[8] A promotional masterpiece with a coordinated network of high-profile social connections became a public relations nightmare. When you mismanage Fuels, you expedite the process of reaching their limits and become an impediment to the spread of your own fire.

MANAGING SELF-MISMANAGEMENT

We all have irrational moments. You may choose the worse option when you know what's right. You stay up late when you know you have to get up early. You have one more drink because you "feel fine." You spend money you don't have on brand-name products even when generics are comparable. You are not a robot. You're not programmed to act rationally all the time. You make mistakes in judgment.

It is very unlikely that any one mistake will be the downfall of a lifetime of successes. The key is preventing yourself from reliving the same mistakes over and over. Being human means erring, but it also means having the capacity to evaluate situations and self-correct. However, just having this capability doesn't make it happen. You have to choose it. It's not always the easy choice. Mismanaging your resources can be fun, right?

When your outcomes don't match your expectations, what do you do? Blame others? Accept it as unchangeable? Evaluate what you could have done differently? It's no surprise that the third option is the choice Firestarters make. Stuff happens. It's what happens inside your head after the stuff happens that distinguishes successful Firestarters from those whose flames get snuffed.

SPARK YOUR THINKING

1. Do you have difficulty regulating an area of your life?
2. Is there a choice you can make or a process you can put in place to gain control over these situations?
3. What role does suppression play in your own self-mismanagement?
4. What Fuels are you most likely to mismanage? How can you prevent that?

Chapter 25

PUNISHERS: DON'T LET OTHERS HURT YOU

I hope that people learn from my mistake, and I hope that the fans forgive me.
—Rafael Palmeiro, Major League Baseball player after a failed test for banned substances

In April 2017, Fox News announced that *The O'Reilly Factor* would be canceled in the midst of fallout involving sexual harassment claims against its host.[1] The *New York Times* reported that Bill O'Reilly and Fox News allegedly paid out $13 million over the course of several years to quiet allegations of misconduct.[2] His advertisers fled, and Fox News made the decision to drop the host despite twenty years at the top of the ratings. O'Reilly alluded to a more nefarious motivation in a statement addressing his departure, referring to the "unfortunate reality many of us in the public eye must live with today."[3]

It's difficult for any brand to weather strong allegations of sexual harassment, especially when multimillion-dollar payments are involved. The presence of the payoff creates confusion, uncertainty, and a perception of guilt. Additionally, it is a situation with which almost all of us cannot identify. How many people have you paid not to talk about alleged actions in your past? Likely, none. . . . So you wouldn't understand the motivation of an innocent person paying millions of dollars to accusers if the case wasn't grounded in some facts.

Punishers represent negative consequences for using Igniters or tapping Fuels. Sometimes punishment is legitimate. Other times it serves as a psychological control mechanism. Punishers can threaten your fire, whether you view them as deserved or just a by-product of your situation.

LEGITIMATE PUNISHMENT

Some forms of punishment are viewed as acceptable risks. For example, professional athletes who use performance-enhancing drugs (PEDs) find a way to circumvent rules and expand their influence on their game and their paydays. If they are caught, however, the weight of the enforcers for their respective sports comes down on them in the form of suspensions, expulsions, and stripping of records. Many retired Major League Baseball players whose stats would have previously ensured enshrinement in Cooperstown found that the cloud of PED use in the game limited their case to Hall of Fame voters.

Punishers serve to deter action. When you choose a course, something negative happens. This negativity in turn squashes your motivation to act similarly in the future. If the punisher eliminates a bad behavior (e.g., PED use), it can be constructive. However, punishers aren't always paired selectively. Sometimes, good motivations and positive fires can be stamped out as well.

For example, people who engage in political or social cause protests are often subject to punishment by those in power. The government and many in the society would consider their actions criminal and punish them accordingly. Some of the great Firestarters in history like Susan B. Anthony and Nelson Mandela rose above the punishers. But many others have their fires stamped out by imprisonment and even death. They are often the ones that history has forgotten because their fire was put out too soon.

PUNISHMENT AS PSYCHOLOGICAL CONTROL

Those attempting to control your behavior often associate the undesired behavior with some form of negative consequence. Parents do this. They limit leisure activities, encourage attention to schooling, and direct children's focus to desired areas. When children stray, the result is often some form of punishment. "You're grounded." "You're disinherited."

These reactions are emotionally punishing, and when they are experienced, they serve to dissuade action. For Firestarters, this could mean dissuading positive action. Other people's reactions are strong motivators, especially when those people have some level of control over your outcomes. Your parents. Your significant other. Your boss. All of these people have an influence that likely outweighs that of others in your life.

Punishment can be a nefarious form of power. Imagine you are on your way to dinner with a friend. Almost there, you spot an elderly gentleman who has fallen down. You stop to help him get steady and walk him to a safer place.

By the time you get to dinner, you are twenty minutes late. Your friend looks at you and taps his watch. He says, "I don't understand why you do this. You're always late." You explain the situation, but his negative tone remains. "You could have at least called and let me know." You end up confused because you thought you did the right thing. However, you feel terrible for annoying your friend.

The next time you make an appointment with the friend, you are less likely to allow distractions, even socially appropriate ones, to pull your attention. Why? Because his reaction was a small form of punishment. Punishers are effective because they create feelings of shame, guilt, and disappointment (warranted or not) that you do not wish to experience again.

Some readers may say, "I can't imagine a friend acting like that. I'd tell him to take a hike!" In that case, your words would serve as a punishment for his crassness. He may think twice before criticizing your pro-social actions again. Anyone can serve as a punisher for someone else's actions.

PUNISHERS AND SUPPRESSING POSITIVE ACTION

Nobody likes being punished, but Firestarters are willing to risk it more than most. Innovators risk scorn for a new invention. Instigators risk unintended consequences when they apply scorched-earth tactics to promote change. Initiators face the forlorn looks of their children and spouses when they explain that attention is needed on this or that pet project in lieu of family time.

Punishers arise directly out of actions that violate others' expectations. In the case of antisocial or illegal actions, punishers are justified. This is the foundation of a criminal justice system. You do the crime, you pay the time.

Other times, the meting out of punishment is based on idiosyncratic or random circumstances. These types of punishers are harder to evaluate. If your friend freaks out about your lateness, you must consider, *Was the reason for my lateness important enough to bring about this reaction?*

For Firestarters, the fire is the driving force of action. Punishers threaten to upend their motivation. Remaining resolute when you face known punishers is difficult. But it is a choice. Sometimes, it's easier to give up or change course rather than directly face whatever unpleasant reality awaits.

SPARK YOUR THINKING

1. What punishers in your life prevent you from igniting?
2. What punishers deplete your Fuel?
3. How can you recognize when you are in a punishing situation?
4. What strategies can you use to prevent punishers from having control of your actions?
5. Can you differentiate between legitimate and illegitimate punishers?

Chapter 26

INEPTITUDE: DON'T BE STUPID

I claim to be a simple individual liable to err like any other fellow mortal. I own, however, that I have humility enough to confess my errors and to retrace my steps.

—Mohandas (Mahatma) Gandhi

In the 1986 World Series, Bill Buckner of the Boston Red Sox committed an error that lives in infamy. With the Red Sox up three games to two on the Mets, the momentum seemed to be on their side. Game six went to extra innings. Buckner was playing first base with the game tied and the winning run on second. The batter Mookie Wilson hit a slow roller toward Buckner, but amazingly the ball rolled past Buckner's glove right between his legs.[1] The Mets won the game and eventually the championship.

Was Buckner a bad baseball player? Of course not. He was Major League caliber. But he did suffer through a moment of ineptitude. He committed an error reminiscent of what you'd see at a typical Little League game.

Buckner's moment highlights the dichotomy of ineptitude. It can be either a pervasive skill deficiency or a momentary lapse. For example, some people will never excel as musicians. No matter what instrument they try, it will always sound like a wailing elephant. But even good musicians can play bum notes. A top trumpet player who has the wrong timing could momentarily throw off an entire orchestra. Novices and experts alike can fail due to ineptitude.

INEPTITUDE AND FAILURE

Some people reading the opening of this section would have a hard time equating Bill Buckner's error with ineptitude. Mistakes happen. Failure happens. It's a fact of life. Can't we give Bill a break? Certainly, but you don't learn from breaks. You learn from breakdowns—reviewing them, understanding them, and correcting the actions that led to them. When we discussed mastery mindset in chapter 6, we highlighted Albert Bandura's assertions that highly efficacious or confident people blame failure on lack of effort. In contrast, those without confidence blame failure on their own ineptitude. Even if failure is accidental or serendipitous, people with low confidence blame their abilities.

Can you see how this would start a crazy cycle? Ineptitude, even if it's only perceived, is a killer of effectiveness. People generally don't seek out opportunities where they feel they have failed or think they will fail.

INEPTITUDE AND FUELS

Having access to Fuels doesn't mean you know how to use them appropriately. Power is abused. Opportunity is wasted. Money is overdrafted. You can make an uninformed bet. You can annoy social connections by involving them in nightmares like the Fyre Festival, which we mentioned in chapter 24.

We aren't born knowing how to use every Fuel at our disposal. Think of the novice grill master who overloads the grill with charcoal or the camper who uses an entire bottle of lighter fluid on the first day of a weeklong trip. A more knowledgeable person in the same situations would be more frugal with the Fuels at her disposal.

Firestarters are not immune. People driven by passion, talent, or a mastery mindset may have the right mentality for success but lack the practical know-how. Their use of some Fuels at their disposal is based more on trial and error than having a practical understanding. Sometimes, trial and error works out. You meet new people, hand out a business card with a title of CEO, and it draws admiration. Other times, you misjudge the power of the move. For example, handing a card with a high-level corporate title to a leader of a union in the midst of a labor dispute may result in a different set of perceptions.

Ineptitude is unavoidable in some situations. Tapping Fuels as a Firestarter is a double-edged sword. When you use them right, the impact is exponential. When you waste them, you must restart and reenergize. When you use Fuels ineptly, you also waste time. You must seek out new Fuels and attempt to relaunch. Hopefully, you will learn something along the way.

RECOGNIZING YOUR OWN INEPTITUDE

When others do something stupid, it's easy to see, isn't it? It's as if you're in a theater, the film is moving in slow motion, and you can see the obstacle coming a mile away. Your own ineptitude is harder to recognize. Why? Because we don't like to admit when we act stupidly. We blame others, avoid looking into people's eyes, or try to cover up our mistakes.

Think of the stereotypical nerd who is book smart but doesn't have common sense. This example resonates because we all have an inner nerd. There's something at which you excel. Conversely, there is a certain area in your life where you always act stupidly. Do you have a soft spot for sales pitches? Do you have a family member who always pulls you into bad situations? Do you try to do your own taxes without knowing anything about accounting?

We all do things we're not good at doing. This is a necessity. You can't be good at everything. So how do you avoid the pitfalls of these situations? Ask for help. Humble yourself. Recognize your own fallibility, learn from it, and always remember to help others when you're watching their slow-motion car crash. You never know when you'll be in a similar situation.

SPARK YOUR THINKING

1. Do you believe that most of the mistakes you make are because of ineptitude or other factors?
2. When you feel inept about something, what strategy do you use to fix the situation?
3. How do you determine if something is a pervasive skill deficiency or a momentary lapse?
4. When you are working with others, how do you handle ineptitude? Is your strategy working for you?

Section 6

FIRESTARTERS TELL THEIR STORIES

The Firestarter Framework

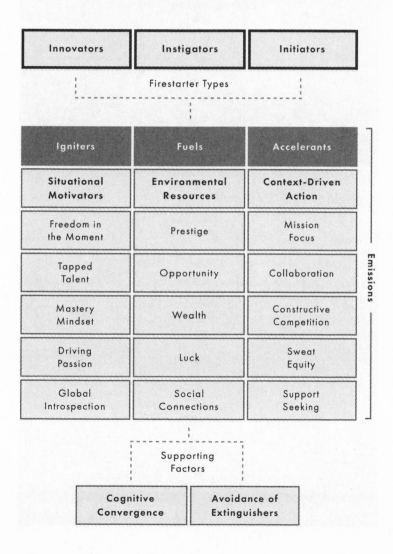

Chapter 27

INNOVATORS CREATE THINGS

The moves we make today have to be bigger than the moves we made yesterday.
—Shaun Lewis, N Group Consulting Services, LLC

Our society has a deep fascination with Innovators. Type the word into Google and you will get almost ten million hits. Type it into an Amazon book search and you will get more than two thousand titles.

Sometimes, Innovators come up with something completely new. Sometimes, they tweak things that already exist. Often they are equated with technological innovation, but Innovators are woven into the total fabric of our world. You can find them in service industries, manufacturing arts, education, politics, and every possible industry. You can find them in rural areas and urban centers. They can be innovating on small or large scales.

Innovators have a mindset. They need to create, to look at new ways of doing things. In this chapter, we share with you the stories of a range of Innovators. Some will be familiar to you; others will be completely new. We talked with them to better understand how they got to where they are and what they perceive to be the difference between Firestarters and others.

They have self-selected themselves as Innovators based on a question we asked them: "Innovators create things. Instigators change things. Initiators begin things. If you had a pie chart, what would your percentages be?"

The people who are included here rated Innovator as their highest percentage. They proudly wear the badge of Innovator. Discover new parts of yourself as you read how these Innovators ignite their lives and the world by creating things:

- John Sculley: Former Apple CEO Disrupts Markets for Noble Causes
- Maya Penn: Oprah SuperSoul Influencer Discovers Inspiration = Youth + Creativity + Business

- Adam Sobel: Executive Chef Conquers Restaurant World
- Dr. Pernessa Seele: *Time* Magazine's Most Influential Person Improves Health with Faith and Education
- Dr. Kirk Borne: Passion Turns Scientist into Number One Big Data Influencer
- Courtney Scott: Top Travel Blogger Finds Freedom While Creating an Industry
- Mindy Meads: Former Land's End CEO Ignites Fashion Brands through Pragmatic Creativity
- Louis Lautman: Movie Producer Extraordinaire Makes Happiness His Mission
- Dr. Barbara Hutchinson: Association of Black Cardiologists President Puts Heart into Entrepreneurship
- Jerrie Ueberle: Innovative Fireball Helps Women Follow Their Purpose Globally
- David A. Fields: Consultant Guru Maximizes ROI and Value
- Dame Shellie Hunt: Lifetime Achievement Awardee Transforms People to Design Their Success
- Rodney Adkins: IBM's "$18 Billion Man" Fulfills Dreams through Focus, Discipline, and Diversity

JOHN SCULLEY

Former Apple CEO Disrupts Markets for Noble Causes

> *I'm not a believer in legacies. I don't spend much time thinking about tombstones.*
>
> —John Sculley[1]

John Sculley is about as close to a household name as a businessman can get. One of the foremost leaders in disruptive marketing strategies, he is a game-changer who took Apple from $800 million to $8 billion, introduced the Pepsi Challenge, and has been involved in numerous successful high-tech companies.[2]

What drives the man who considers himself to be 70 percent Innovator and 30 percent Instigator? "I'm completely absorbed in doing things that are going to make a difference in people's lives," John says. "I have little interest in legacy or any of those things."

To that end, he is currently chief marketing officer for RxAdvance, whose mission is to transform the complex healthcare ecosystem through game-changing business and revenue models. "It's a noble cause. If we can save enough money, there is really enough there to take care of everyone."

In 2013, he joined a team of notable physician, healthcare, and technology entrepreneurs with track records in building companies that disrupt traditional paradigms in healthcare. RxAdvance is a solution with the potential to remove $450 billion a year in drug-impacted medical costs by managing the entire care continuum.

In addition to the financial impact, John is excited about how cloud computing and machine learning can improve the quality of life for the chronically ill. "They typically have eight or nine chronic diseases and take fifteen to twenty-five pills a day. This is the first real solution that reduces drug duplication, interaction, and overmedication."

John sees himself as a designer and builder who solves problems in ways they haven't been solved before. "Everything I work on is about transformation. It's from the standpoint of not blowing up what's there, but basically eclipsing it and making it almost irrelevant."

He believes that companies that focus on how to take competitors down are taking the wrong approach. "I look at why an industry is the way it is and how it defines success. Often you find that the success metrics of industries become outdated."

Solving problems that achieve more, make customers more satisfied, and completely change the economics of an industry is what he calls "adaptive innovation."

He cites Tesla's higher market value than Ford as an example. "Who would have guessed that ten years ago a start-up could come into the same industry and be worth more than one of the market leaders?"

That's when he starts talking about Darwin. "If you go back and read what Darwin wrote, he never talked about the survival of the fittest. He talked about the survival of those species that are able to adapt to the changing environment in which they lived. Some adapt and others die. But it's not about a battle; it's not about one dinosaur fighting another dinosaur. It's about being able to adapt to a changing environment."

His own start as a Firestarter was forged as a teenager with a bad speech impediment. He remembers one teacher in particular: "He accused me of being stupid because I couldn't answer a question he asked. I knew the answer to the question. I just couldn't say it."

He eventually overcame his speech problem when he realized it would limit him from doing what he wanted in life. With his usual thoroughness, he approached the challenge through research and action. He also sought out expert help—another strategy he has employed throughout his career.

Now an accomplished public speaker and television personality, his ability to overcome obstacles may be at his essence as a Firestarter. "Whatever obstacles are in front of you, you get those obstacles out of the way."

Why Firestarters Are Different

People who transform the world are the ones who look at the same facts that everyone else looks at, but they interpret them entirely differently.

John spends most of his time with people who are actually accomplishing their dreams, not just thinking about them. This has led him to several observations about what makes Firestarters different than other people.

First is insatiable curiosity with the ability to frame facts in a different way. "You have to be the kind of person who observes things around them and wonders how they work and if there is a better way."

He recalls taking long walks with Steve Jobs where Jobs would engage in what he called "zooming out and zooming in." Zooming out was looking beyond the things you were actually working on to other places and then connecting the dots. For example, Jobs studied calligraphy before he started Apple and visited the Palo Alto Research Center where he saw $80,000 engineering workstations with graphic interfaces. Instead of making computers more powerful, he decided to focus on computers that were easier and inexpensive for nontechnical, creative people to use—a totally new idea. Then he zoomed in to figure out how to simplify it to make it easy for customers.

That leads to John's second observation: "It's the customer plan that counts. It's not the business plan. Business plans are usually a resource allocation exercise. The customer plan comes from a totally different place. It starts with saying, 'What's a really big customer problem we can solve? What's in it for the customer?' And then at how you make that customer incredibly happy."

When he and Jobs worked together, they always started with the customer's experience. "At that time in Silicon Valley, everyone was an engineer, and the engineers always started with the technology. How do we make it faster? How do we make it bigger? They weren't starting with the user experience. We were designers, not engineers; we saw things from a different perspective."

Finally, John believes Firestarters frame things within the context of a noble cause. "If something has an overarching mission that excites people, then it will attract really talented people to the team. That creates a momentum of its own, well beyond whatever the invested resources are that go into the company."

He remembers being in an engineering lab with Bill Gates and Jobs the year before the Macintosh went to market. The conversation was about their noble cause and how they were going to change the world—one person at a time—by empowering knowledge workers with tools for the mind.

"My past experience was all about somebody wins, somebody loses," John says. "Listening to these two guys talk business was in a context I'd never heard before. A noble cause. It stuck with me all of these years, which is why I'm so focused on a noble cause [that] potentially could take hundreds of billions of dollars cost out of the US healthcare system."

MAYA PENN

Oprah SuperSoul Influencer Discovers Inspiration = Youth + Creativity + Business

> *Everyone has the power to make a difference no matter how young or how old you are. I think that's a message that everyone needs to understand.*
>
> —Maya Penn[3]

It is easy to be humbled by a teenager who started her eco-friendly fashion company when she was eight; has almost two million views of her TED Talks; was chosen by Oprah Winfrey as one of her SuperSoul one hundred influencers; and made history by creating an animation for the first-ever digital report presented to Congress to advocate for a women's history museum in Washington, DC.

Oh, did we mention that she is also the author of *You Got This!: Unleash Your Awesomeness, Find Your Path, and Change Your World*; is producing a new environmentally driven animated series for kids, *The Pollinators*; has been featured by pretty much every major media outlet; formed her own nonprofit; partners with Google to engage girls about coding; and was selected by Magic Johnson for his 32 Under 32 series?

But Maya is just getting started. At seventeen, this Firestarter, who considers herself 50 percent Innovator, 30 percent Instigator, and 20 percent Initiator, cannot be contained. "I've had a huge opportunity to explore all my ideas, follow my creativity and curiosity, and try them out without lots of second thoughts or opinions," she says.

Maya attributes this freedom to her creative and entrepreneurial parents, being homeschooled, and being inspired by people she has met on her journey. "What makes someone an influencer to me is not only their experiences, but their positive energy. I think it's so important to surround yourself with people that want to uplift you and want the best for you."

She also is aware that her age has been a huge factor in her career to date. "I was eight when I started, and I was serious. Some people, however, think you're playing pretend and don't want to help you. But age has also opened a lot of interesting doors for me, because it is very unique for someone my age to do all these different things that I do."

Maya loves creating new things and being on the cutting edge of ideas that can help change the world. When she started her company in 2008, neither social entrepreneurship nor eco-friendly fashion was a common term. She also has been able to merge innovation with instigation.

One of her most intriguing social ventures helps young women in third world countries stay in school. Maya found out that many girls lose crucial time from school because they don't have sanitary pads to wear during their menstrual cycle and must stay home. She created eco-friendly and reusable sanitary pads that could withstand hundreds of washings. Now she is partnered with organizations that are distributing the pads to Haiti, Senegal, and Somalia among other places.

"It's been so incredible to see. I'm just really glad I'm able to make an impact on girls' and women's lives just by something we take for granted here because it's so easily accessible to us," Maya adds.

What Makes Her a Firestarter

Knowing that I've had the power even as an eight-year-old to help inspire others to make a difference is really incredible to me.

Maya believes that the traits that have enabled her to be a Firestarter—creativity, passion, and fearlessness—are traits that can be cultivated in all of us. One of the reasons she wrote her book was to give teens and young adults a blueprint to develop their own power.

"Sometimes," she says, "it's a little daunting to think that I'm a leader and role model especially for kids or younger teens. But I'm definitely happy and excited that I'm able to inspire and shape the upcoming generation just from a simple idea that I had."

Her passion is what keeps her going. "If I ever feel like I'm at a roadblock and I don't know how to proceed, I just remember why I started. I wanted to do something that was really a creative outlet for me but also made a difference and got results."

A strong analytical side balances her passion, and her creativity is powered by her ability to focus and figure out how to make ideas tangible. "You need to check in with yourself and figure out what are the priorities. If you have an idea that doesn't fit at the time, it's important that you save it in the idea bank because you never know when it may come in handy."

Why Firestarters Are Different

> *Everyone has the power to make a difference no matter how young or how old you are. I think that's a message that everyone needs to understand.*

Firestarters are different for two reasons, according to Maya. First, they have discovered what they are passionate about. Secondly, Firestarters get over the "Can I do it?" hump. "I think there are a lot of people out there who are not sure what it is they really care about. And a lot of people don't realize that it's a really big leap from what they care about to doing something about it."

People who are not Firestarters can get overwhelmed or believe that someone more qualified than them can make it happen. "They don't realize they can be the change they want to see in the world. This is prevalent among young people because you can put a limit on yourself because of your age."

As she travels the world, Maya inspires many people and is an inspiration to so many others. She tells the story of ten-year-old Saria who emailed her after reading her book and then met her in person: "She told me how she was inspired to give back and start her own projects to help other people. It was really cool to see my eight-year-old self in another girl who is excited and wants to make an impact. It's really cool to see the next generation of innovators, activists, and entrepreneurs."

ADAM SOBEL

Executive Chef Conquers Restaurant World

> *To be a great chef, you have to be able to connect with guests,*
> *want to take care of people, and wow them with food, hospi-*
> *tality, or humor.*
>
> —Adam Sobel[4]

Adam Sobel grew up in a close-knit, high-energy family with a Russian Jewish grandfather, who loved adventures, and an Italian Catholic grandmother, who loved to cook. At age four, he was already in the kitchen. "My grandmother cooked delicious rustic Italian food, but the thing that made her really special was the love she put into it and how much joy she got from making everyone happy," Adam says.

A poor student who had difficulty focusing in the classroom, he enrolled in vocational classes for culinary arts at age fifteen, and his life changed. "Every person has something they can be great at, but not everyone has the opportunity or is drawn to the light. The first day I was in class I knew that it was going to be my career. It was really incredible actually. Divine intervention."

The 40 percent Innovator, 40 percent Instigator, and 20 percent Initiator went on to build his credentials, studying at the Culinary Institute of America with chef positions at Ogden's, Guy Savoy, and RM Seafood in Las Vegas; Bourbon Steak in Washington, DC; and RN74 in San Francisco. He worked for some of the leading chefs in the world: Bradley Ogden, Guy Savoy, and Rick Moonen, achieving high accolades along the way such as being crowned the King of Porc at the prestigious Grand Cochon and appearing as a guest judge on Food Network's *Chopped*.

Today Adam develops restaurant concepts and drives the cuisine for the MINA Group where he works closely with founder and CEO Michael Mina, a celebrity chef, restaurateur, and cookbook author. "Michael and I have been very close for a long time, but to get a partnership in a company like this took a lot of sacrifice and a lot of hard work. I had to earn his deep trust."

Adam feels he has made a significant contribution to the success of his company. "It all comes down to working really hard for the right people with the right mentors. The doors open, and you build relationships, make delicious food, be innovative, and get acknowledged for it. It just keeps stacking the deck. Now we're to the point where we are able to kind of write our own ticket."

What Adam loves about his work is that it never gets boring. "I have to wear many different hats. It's always evolving and changing with new partners and concepts. The food world is evolving. I feel like I have an opportunity to create every day, and that is really important to me."

To that end, he constantly has new projects to keep him busy. "I work on ways to be different and stand out—to lead the pack." There's the fast casual restaurant called Adam's Nana Lu, which pays homage to his grandmother's Sicilian specialties. There's Cal Mare, a fine dining concept featuring coastal Italian Cuisine. And there's Lafa with its Middle East meets San Francisco cuisine.

Connections are one of the key factors in making him a Firestarter. He attributes a large part of this ability to his personality, humor, and being open. "I've been able to connect with the great people in the business; the most innovative chefs and winemakers in the world have helped me learn and opened doors."

He also believes that luck plays a factor, although it has its own variation in the restaurant industry. "In other businesses, some people get lucky right out of the gate. They have a great product and are lucky to connect with someone. In my business, it takes years for all the dominos to fall into place. Then wealth comes toward the end."

Adam needs to be inspired to keep his passion burning. He goes on food trips around the world regularly to experience cuisine and meet new people. "You can't sit on your hands and wait for things to come to you. When you're a chef, you have to read. You need to be a student of what's going on. You need to have open eyes, open ears, and be open to meeting people, talking, and doing some digging."

Recently, for example, he went to Italy and made mozzarella with the greatest cheese maker in the country. "That only happened because we sought it out. It wasn't like I was going to sit home, watch TV, and then be inspired by something I saw. You have to go and do it."

Why Firestarters Are Different

The passion that is within me drives me every day.

Adam believes that people who haven't ignited their Firestarter potential have not found their calling. "I feel like I'm operating on high octane fuel. Someone who doesn't have that passion or hasn't found their calling does not have that pep in their step, zest for life, and true happiness."

But passion is only one component of what he thinks makes Firestarters in his industry. "I really love to cook, make delicious food, and create amazing teams. Then it's about getting acknowledged for excellence. That drives us also because with that comes more opportunity. We also do what we do because we want to make money. We want to be successful. We want to have financial freedom one day."

He also believes it is about having a mission. "I feel being a chef influences what people eat and how they eat. I'm able to contribute to my communities. I'm able to help aid sustainability and the future of food in our country. And every chef has an ego, so I can't deny that it's definitely about building my legacy. It's always been about how I'm able to impact and make a difference."

DR. PERNESSA C. SEELE

Time Magazine's Most Influential Person Improves Health with Faith and Education

> *Our faith lights the way, but creativity sustains change.*
> —Dr. Pernessa Seele[5]

Pernessa Seele wanted to be Aretha Franklin when she was growing up. She did not accomplish that dream, but she did earn the R-E-S-P-E-C-T of political, religious, healthcare, and community leaders throughout the world.

Best known as an AIDS activist and now as a health activist, Pernessa created the National Week of Prayer for the Healing of AIDS thirty years ago. More than ten thousand religious organizations participate in this effort.

Sixty percent Innovator, 20 percent Instigator, and 20 percent Initiator, she has earned accolades including being named as one of the most influential persons in the world in 2006 by *Time* magazine. She's an author of multiple publications, and her latest is *Stand up to Stigma: How We Reject Fear and Shame*.

Pernessa grew up in Lincolnville, South Carolina, a town formed by ex-slaves. Her great-grandfather was one of the founders. The church was at the center of her life, and all pastors were powerful leaders in the community. "If you were sick, the pastor was called first," Pernessa says. "If you died, the pastor was called first. And the people in the church responded to every single crisis in the community big or small."

She joined Harlem Hospital at a time when 100 percent of the people who came into the hospital with AIDS died. What she experienced there changed her life. "People were dying alone." It was inconceivable to her that spiritual people and churches would not step up to the plate.

With only three days on the job and not a single contact among the pastors, she began mobilizing the faith community by holding a Harlem Week of Prayer for the Healing of AIDS, a weeklong HIV education campaign that gave birth to the Balm in Gilead. Her international organization is celebrating almost three decades of service and provides technical support to more than twenty thousand faith institutions regarding the implementation of health education and service programs.

While Pernessa's reputation is that of a change agent, she believes that it is innovation that is her primary strength as a Firestarter. For example, coming up with the idea for the Harlem Week of Prayer grew into the National Week of Prayer for the Healing of AIDS, which then served as a model for engaging faith communities to address HIV and AIDS globally.

She was innovative in linking the role of faith to HIV. She launched Our Faith Lights the Way, a national faith-based HIV testing campaign, which has been adapted for use in Africa. Faith-based HIV testing programs changed CDC and public health departments.

"These institutions began to realize that people came back to the church to get tested and to get their results when they would not go to the health department. So, city and state public health departments began funding churches to do HIV testing because they were getting better responses. Totally surprising, CDC launched an HIV faith initiative."

She used her ideas to mobilize the African American community in other ways. In 1996 while driving to work, she thought it would be great to have opera singer Jessye Norman support an AIDS benefit in Harlem. Norman said yes and invited Whoopi Goldberg, Maya Angelou, Elton John, Bill T. Jones, Max Roach, Anna Deavere Smith, and Toni Morrison to join her for the first major, star-studded AIDS benefit focusing on the plight of HIV among African Americans.

Then she created another first for the African American community when Barack Obama was elected president. "I had never been to an inaugural ball before, but I knew we needed to [have] an event to acknowledge the efforts of many that got us to this place in history. So we created the African American Church Inaugural Ball and honored Desmond Tutu, Colin Powell, Jesse Jackson Sr., Myrlie Evers-Williams, Rev. Joseph Lowery, Bishop Barbara Harris, Rev. Dr. Gardner C. Taylor, Maya Angelou, and others."

Today she continues her work and has expanded her mission. "Every church serving African Americans needs to have a health ministry because we have among the highest rates of diabetes, more Alzheimer's, and more HIV. I want to help African American churches become more relevant around health. We must become healthier people in mind, body, and spirit."

Why Firestarters Are Different

> *I was raised by old black women who taught me how to trust the voice of God. Firestarters believe in that voice within us.*

Pernessa believes that many Firestarters are born with a passion to make things different. Others are placed in situations that ignite that passion, and they will do anything to make it happen. "When I started the Harlem Week of Prayer for the Healing of AIDS, nobody knew what I was talking about. I wasn't even too sure what I was talking about. I had absolutely no money, only a 'little' idea, and I trusted the source of that idea."

She says believing in what you believe in—even if it doesn't make any sense—is one thing that makes Firestarters different. "When you believe that you're called to do something, nothing is going to stop you because you believe it in your heart, soul, and mind. You are going to get it done no matter what."

DR. KIRK BORNE

Passion Turns Scientist into Number One Big Data Influencer

> *I was very interested in what was going on in the rest of the world around data mining, business intelligence, and data visualization, even before we had terms like big data, data science, and analytics.*
>
> —Dr. Kirk Borne[6]

If you want to find a person who epitomizes the expression "scary smart," you don't need to look further than Dr. Kirk Borne. Data scientist. Astrophysicist. Space scientist. He is considered one of the rock stars of big data with a Twitter following of more than 150,000 people. He is ranked at the top in influencer rankings for big data, data mining, machine learning, data science, and Internet of Things categories.

Growing up in an air force family, the man who categorizes himself as 80 percent Innovator and 20 percent Initiator was always exploring new and different things as they moved around. At age nine, Kirk decided that he wanted to be an astronomer. His passion for understanding the world was reinforced by people like a middle school math teacher who was a retired colonel. She instilled discipline in him and opened up the world of learning science and math.

His talent and abilities blossomed in high school where he ended up ranking at the highest level in national math competitions and being the only person to ever get a perfect score on a national chemistry competition.

"I never really thought of myself as being a really gifted person because I had mediocre grades in grade school," Kirk says. "In high school, all of a sudden I had universities inviting me to apply for scholarships. It was an awakening to what was already in me—the ability, aptitude, and desire to pursue science as a lifelong passion."

He earned his doctorate in astronomy from Caltech and then spent eighteen years at NASA where he worked with large data sets that came from satellites. His curiosity drove him to dive deep into computer publications for any discussion about data mining and business intelligence and to keep his own personal log of related articles. He also moved into the educational sector where he helped start the world's first undergraduate data science program at George Mason University in 2007.

In 2013, the top consulting firm McKinsey reported about the talent shortage in big data analysis,[7] the White House announced a National Big Data Research and Development Initiative,[8] and *Harvard Business Review* named data scientist as being the sexiest job of the twenty-first century.[9] And Kirk discovered Twitter. No longer able to keep up with his article log, he started using Twitter for his personal benefit to keep a curated log of articles, conventions, start-ups, and discoveries in the field. "A year later, one of my Twitter connections asked me how it felt to be number two in the world of big data influencers. I didn't know what that meant. Then I realized that what I was doing really resonated with a lot of people and it sort of became my avocation—my life on Twitter."

What Extinguishes His Fire

My usual approach to Discouragers is, 'Okay, I know what I am. I have faults. I am an imperfect human being, but I am going to go forward anyway.' What can I do? Sit here and wallow in it?

Kirk is not good at saying no. He has a hard time saying no to learning new things, solving new problems, and taking on new projects. "I think being able to manage the wealth of things that are going on—the kid in the candy store problem—becomes discouraging because I really do like all this stuff. I would like to be able to bring more, which is impossible."

But he has an even harder time taking no for an answer. He recalls instances in his life when he was discouraged from being a manager of people or finding a job in his field. "The Discouragers were external, and until you really had a firm footing on the ground, sometimes those voices carry more weight than they should in your life."

He is particularly passionate about helping students overcome Discouragers. As a teacher, he often had students who said they hated math and science and were only taking his course as a requirement. His response would be to make it fun and interesting. He even taught calculus in his data science class and didn't tell the students that it was calculus until the end of the course.

Kirk says, "One of my most gratifying moments was when one guy came up to me after the final exam, shook my hand, and told me that even though he had hated science his entire life, this was the best course he had ever taken. It brought tears to my eyes."

Why Firestarters Are Different

> *It does not really matter where it comes from, but there has to be something that wakes it up inside you to say yes I can do this. It is valuable, and I can do it.*

Firestarters believe in themselves, according to Kirk. Sometimes they are people who self-ignite. Other times, they need someone to ignite the spark in them.

"Once that passion starts burning, it is hard to not do something," he says. "There was one student that I taught who was a medical technology major but discovered a passion for data science. So he minored in our data science program. After graduation, he sent us a letter [explaining] that he had gotten his dream job because he has minored in data science and none of the candidates even knew what it was."

That goes to prove another concept that ignites Kirk's passion: "If you are doing what you love, you will always love what you are doing."

COURTNEY SCOTT

Top Travel Blogger Finds Freedom While Creating an Industry

> *I get notes from women and people all over the world asking how they can follow in my footsteps. My mission is clear—to make this lifestyle accessible to other people.*
>
> —Courtney Scott[10]

Courtney Scott was two years old growing up on Long Island when her dad left. Her mother supported the family, doing everything from scrubbing toilets to babysitting while working three jobs and going back to school so she could eventually become a school principal.

Courtney says, "My mother taught me two very important lessons—the power of believing in a dream and the necessity of an unwavering work ethic in order to make that dream happen. There were a lot of starts and stops in my path, and I always had my mother as a beacon to remind me that it's possible."

The other gift her mother gave her was a video camera. "I just started documenting everything. I was directing elaborate shows in my basement with my best friend Stacey. I loved the creation process."

In 2007, the woman who categorizes herself as 50 percent Instigator and 50 percent Innovator left her New York City marketing job and booked a one-way ticket to Italy where she began to teach English and create travel content. Her timing was impeccable. In comparison to today, few women were living abroad and creating blogs. In fact, travel blogging was in its infancy.

"I had a video camera at all times, and the concept of being able to be a freelance travel filmmaker started to come into vision. The industry was being created at the same time as my dream was being built. I was fully inspired, super creative, writing about all these experiences I was having and getting paid very little."

Today, Courtney is a travel filmmaker, TV host, blogger, and leading on-air travel expert. She is a regular contributor to CNN, ABC News, and the *Today Show*. Her visceral style of storytelling has been seen in brand campaigns with Porsche, Expedia, Refinery 29, Vanity Fair, Disney, Avianca, and Marriott as well as destination partners like the Galapagos Islands, Cuba, the Hawaiian Islands, and Colombia. She is passionate about building community within the travel industry and is a frequent speaker at travel events and summits like Women's Travel Fest.

What Makes Her a Firestarter

> *Once you free up the space in your mind and in your life, it really ignites the fire of your creativity. It opens your heart.*

Freedom, opportunity, and hard work are factors that make Courtney a Firestarter. "Freedom allows my passion to sing. Being able to create projects that are pure passion and collaborating with really fascinating people ignites my fire."

The opportunity to have new experiences fuels her by opening up her mind and connecting her with different people around the world. "There is always something lurking around the next corner that I haven't discovered. I know that we have one life; it's short, and there are so many places to see, so many experiences to have."

She also acknowledges that hard work pays off as well as a bit of competitiveness. "I've certainly built my business on pure . . . blood, sweat, and tears, but I also think a little bit of competitiveness is healthy. I don't like to do what everyone else is doing. In our industry, you aren't getting a paycheck week to week. You really have to work for your success. Part of that success is staying relevant and creating a unique style of content."

What Extinguishes Her Fire

> *It's important to stay traveling, but I have to take a step back and decide how to make it all happen—how to have a family and a future while still being this digital nomad.*

Nothing excites Courtney more than traveling, being adventurous, and using all her resources to move forward. "The industry I'm in fuels the creation process. If I am not traveling and creating content from my travels, I lose relevancy."

At the same time, at age thirty-five, she is trying to put down roots and build a future. "This is a common theme when I talk to other people who are digital nomads now in their thirties. I have to temper things a little bit and be mindful, but it's completely possible to do it all."

Why Firestarters Are Different

Once your fears dissolve and you dream bravely and dream consistently, I think that is the ticket to finding out what you're truly destined to do.

Courtney believes that a simple four-letter word separates Firestarters from others. Fear. "Fear is crippling. Fear is confusing and distracting. It can be so powerful in a person's life and cloud your real judgment and what it is that you're destined to do in your life."

Courtney says people numb themselves to their day-to-day existence, not truly recognizing and awakening to the power they have. "Some people never get to the point where they feel powerful enough to explore what it is that truly makes them happy."

She views herself as lucky because when she made changes in her life, fear did not enter into the equation. She asked herself what was the worst thing that could happen if she followed her dreams and realized it was not a big deal—a different job or going back home. "Fear, anticipation, and projecting all that could happen if you do something drastically different—those are the things that cripple people. Firestarters don't let these get in the way."

MINDY MEADS

Former Land's End CEO Ignites Fashion Brands through Pragmatic Creativity

My secret sauce is knowing the numbers and having the right product at the right time at the right quantity and the right size. It's that precision of excellence that will get your results.
—Mindy Meads[11]

Mindy Meads is a Firestarter in an industry where staying true to your principles is challenging in a world of buyouts and consolidation. Her dream was always to be in retail. "As a teenager, I knew I wanted to be a buyer," she says. "I loved numbers, clothing, and art. I never had dreams of being a CEO. It was one step to the next. I just wanted to do that job well. And then the next one came."

Forty percent Innovator, 30 percent Instigator, and 30 percent Initiator, she has spent most of her stellar career in the fashion mecca, New York City. At Macy's, she worked her way up with a new job every year and ended up as senior vice president of apparel. She then moved to the Limited where she switched from buying clothing lines to creating them.

At Land's End, she helped take the company from $650 million to $2 billion. When Sears bought it, she did not like how employees were treated and strongly disagreed with distributing the brand in Sears and Kmart stores. Ousted from her position as CEO, her concerns eventually proved correct as the once thriving brand encountered trouble.

Still Land's End remains her favorite company because of its culture. "It was all about the people and the customer. The founder used to say that everything else takes care of itself. I was proud of the people and watching them grow."

Using a potent blend of creativity and attention to numbers, Mindy engineered her other major financial success at Aeropostale where sales went from $1.2 to $2.5 billion in two years under her leadership. In addition to the right team, she credits accelerating growth with understanding the customer.

"I have a good ability to really hone in on what the customer wants—the right product, the right quality, and the right price," she says. "I keep it tight with really good execution by identifying an opportunity and understanding how much to buy as well as contingency plans to have enough to hit potential."

Today she has left the intensity of New York City for Sun Valley, Idaho. Unable to keep her hands out of building a fashion brand, she has invested her money, time, and talent in a small activewear company, Vie Active. "In essence, big and small companies are the same. It is all about discipline. And you have to keep it simple. When I was at Land's End, ten thousand employees all had the same objective."

What Extinguishes Her Fire

When I don't have the full team in place, that is a showstopper for me.

At the end of the day, it is all about the team for Mindy. While she has both let go of employees and left companies herself, she remains friends with many of the people she has parted ways with.

"I need to be around people who are all going in the same direction. In some cases, I had to make hard decisions. I have made choices to leave, and I am not afraid of that. You have to be happy at what you are doing. You have to be responsible for making that change yourself."

Why Firestarters Are Different

I am incredibly driven. I have a vision. But I do not do it on my own. I bring other people along.

For Mindy, being a Firestarter is passion, drive, persistence, and, most importantly, leading a great team. "I have been fortunate to be given the right opportunities, and I have been fortunate to make them as good as they can be. Primarily by getting great people to work with me. I am not someone who does it by herself. I am very much a team player. I am good at identifying talent that can come along with me and help me answer all the questions."

LOUIS LAUTMAN

Movie Producer Extraordinaire Makes Happiness His Mission

My mission has shifted and changed over the years, but the reoccurring theme has always been to be happy.
—Louis Lautman[12]

Louis Lautman was five years old when his mom refused to give him money to buy bubblegum. So he went to the kitchen, emptied pretzels into his little red wagon, and walked around the neighborhood selling them. The young boy, who now considers himself 40 percent Innovator, 30 percent Instigator, and 30 percent Initiator, got the bubblegum.

He also got a taste of sales and entrepreneurship that have been the drivers of his career and personal life. By thirty, he had produced *The YES Movie*, an inspirational film in which some of the country's most extraordinary young entrepreneurs tell their secrets to success. He also founded Supreme Outsourcing, a company that has given him the freedom to live and travel anywhere in the world.

Before starting his own companies, he did professional sales, training, and management in New York City. He also went to work for Tony Robbins. "He had a real major impact on me in terms of the mindset that you have to have to be successful," Louis says.

The death of his brother when Louis was twenty-three had him reexamining his life. "I wanted to do something bigger. I started looking into personal development and life success. I wanted to inspire more young people to become entrepreneurs, so I came up with the idea of the Young Entrepreneurs Society, building a community and making a movie to get the message out."

One dominant theme of *The YES Movie* is that business and personal life are synergistic and must support each other in order for people to find happiness. The movie also debunks what Louis feels are myths about how people create a life of prosperity, wealth, and abundance. "I believe that there are 'facts' that are untrue and limit life such as the belief that you need to go to college after high school or that if you work long and hard at your job you will get ahead. Instead, you need to embrace fresh constructs from a new generation of dynamic, highly motivated, and incredibly successful young people of around the globe."

For Louis, it all comes down to two words: be happy. He relates a story about John Lennon that had a profound effect on him. A teacher asked a young Lennon what he wanted to be when he grew up. Lennon said, "Happy." The teacher said Lennon did not understand the assignment. Lennon replied, "I don't think you understand life."

Enjoying what he does every single moment is one of the fundamental factors that make Louis a Firestarter. "I know what I enjoy, and I know what I don't enjoy. It's important to take an industry that you like and quickly figure out what you like to do and what you don't like. That way you structure your company so you can hire someone or figure out a way to get things done."

What Extinguishes His Fire

I'm the pit bull. You push me down, I'm coming back again.

While he considers lack of money an obstacle that can temporarily put out a fire, he also believes it is a motivator to make you work harder. He feels the same about distracters. "When people didn't believe in me, rather than that letting it get me down, I let that fuel my fire."

Nor does he let self-doubt get in the way. "I have found that the quicker I create empowering emotions, the sooner I am going to get there. My father told me a story a long time ago about two neighbors whose houses burnt down. One neighbor sat there crying and angry. The second neighbor immediately started rebuilding. It made sense to me. I'd rather have the house than sit around moping."

DR. BARBARA HUTCHINSON

Association of Black Cardiologists President Puts Heart into Entrepreneurship

> *My goal in life is simple—to prevent people's untimely death from heart disease.*
>
> —Dr. Barbara Hutchinson[13]

Dr. Barbara Hutchinson considers *Blue Ocean Strategy* one of her favorite books and defines herself as 50 percent Innovator, 40 percent Initiator, and 10 percent Instigator. The cardiologist constantly looks for opportunities to provide new services in heart disease awareness, prevention, and treatment.

These opportunities include creating a sleep management practice, a sleep and heart equipment business, innovative transitional care cardiac programs in skilled nursing facilities, and concierge services for at-risk VIPs. Her latest effort will bring cardiovascular care via telemedicine to the little village where she was born in Tobago.

Additionally, Barbara serves as president of the Association of Black Cardiologists and is an active participant in international women's groups like the Women Presidents' Organization. There she works to help those populations at higher risk for heart disease, the number one killer of both men and women.

Barbara explains, "I come across women every day with clear cardiovascular symptoms that have been blown off because the physicians caring for them don't realize women can present differently than men. The result is that women have more significant disease by the time they are taken seriously."

Growing up in a close-knit family, Barbara was highly influenced by her schoolteacher father and nurse mother. "We learned at a very early age not to rely on our smarts or intelligence, but to realize that it was God who gave us the wisdom to be able to do well."

Her desire to become a cardiologist grew out of her curiosity about why her mother's family was healthy and her father's family had more health challenges even though they came from the same area. When told that she needed to get another degree in order to be accepted into a US medical school, she went to Howard University where she earned a PhD in cardiovascular pharmacology. Then she graduated from the University of Maryland in Baltimore where she excelled as the first African American medical school class president and chief resident in internal medicine.

She created her private practice, Chesapeake Cardiac Care, and then looked for other ways to save lives. She learned that obstructive sleep apnea could cause fatigue, high blood pressure, high blood sugar, abnormal heart rhythm, heart attacks, strokes, and sudden death. "I became very proactive in sending patients to have sleep studies done. But I became highly frustrated because the results were never sent back to me. I'd see patients six months later, and they had severe sleep apnea that was never addressed."

So she started a sleep clinic for her current patients and then expanded to a sleep study lab for others when she built out her office space. She became board certified in sleep disorder and instituted a system that took patients from testing to treatment and follow-up. Recently she opened an online company to sell sleep and heart equipment and supplies.

Her Innovator and Initiator engage in full gear as she looks for other solutions to heart disease. For example, she discovered that large numbers of patients are readmitted to hospitals from skilled nursing facilities due to preventable cardiovascular causes. So she created a transitional care cardiac program for at-risk patients. The result has been a drop in readmission rates from double to single digits.

In another venture, she observed that too often business owners and executives die of heart disease because they do not see a cardiologist until it is too late. So she created one of the first independent cardiologist concierge practices where these individuals could receive very private and personal care that would not impact their high-profile jobs.

"My concierge practice is an example of reaching across the aisle," Barbara says. "It's hard to believe that in this day and age educated people would drop dead at forty from a heart attack because they do not want to be seen going into a cardiologist's office. But it happens, and the impact on business, communities, and families is devastating."

What Extinguishes Her Fire

> *Faith is very central to my life because I realize everything that I do and all my abilities come from Him.*

Barbara is not a woman who believes in Extinguishers. In fact, the only one she cites is money, which sometimes restricts her from doing things faster to accelerate her ventures. Losing patients does affect her, but it never extinguishes her fire. She acknowledges that she cannot save everyone, but as long as she provides him or her with state-of-the-art treatment, she is able to move on.

And she relies strongly on faith. "There's not a day that goes by where I don't tell the Lord how much I appreciate the opportunity to help people. When I have a difficult case, I realize who really is the one enabling me. I go to Him and say, 'Look, this one is really tough. Please give me the wisdom and show me what I can do to help this person.'"

Why Firestarters Are Different

I think it is the passion to reach more people and a variety of people.

Wanting to make a difference in as many people's lives across cultures is the factor she believes that distinguishes a Firestarter from others. Because heart disease is a killer that crosses race, culture, gender, and economic status, Barbara ignites her world with multiple ventures that take her beyond the realm of most cardiologists. "You have to be passionate about what you do and who you are doing it for. You have to reach as many people as possible to make a difference."

JERRIE UEBERLE

Innovative Fireball Helps Women Follow Their Purpose Globally

I'm not interested in creating something new that doesn't burst into flames or heat the world up.

—Jerrie Ueberle[14]

At seventy-six years old, Jerrie Ueberle is a fireball. Her journey from the small mining town of Flat River, Missouri—population five thousand—to world changer, educator, mentor, and traveler has ignited thousands of women and men across the globe. With more than 119 trips to China, she draws on a strong sense of purpose to help young women and men answer what is their passion in life and how they can make it happen.

This passion led the 40 percent Innovator, 35 percent Initiator, and 25 percent Instigator to create the World Academy for the Future of Women in Henan Province. Women and men are selected for the World Academy, based on a commitment to advancing women's leadership worldwide. Most of the seven hundred graduates come from poor rural communities and are often the first in their families to go to college.

She is now taking the program to other parts of the world, including opening a World Academy in Nepal where all of the first-year facilitators are women and men who previously volunteered in China.

Originally, Jerrie's career was in speech pathology and audiology. In 1984, she was invited to go to China to speak as a part of a group to Chinese otolaryngologists. She declined the invitation six times and was chagrined when her husband signed her up without her knowledge. When she returned, he wanted to know about her trip.

"I told him our State Department should have taken our passports away," Jerrie says. "We didn't represent our country well. We knew very little about China and less about how they were prepared to address issues in our field. The Chinese are extremely interested in advancing their knowledge, and their curiosity and engagement was key to the exchanges we had. Through these interactions, we learned what we should have already known about their politics, culture, and how they were providing services to hearing- and speech-impaired populations."

This led her to switch careers and create a nonprofit called Global Interactions. With planning partners around the world, the nonprofit tailors study programs, symposiums, forums, and conferences for international counterparts. "If I'd taken more time to think it through, I probably wouldn't have done it. But I believe by connecting people whose professions and careers shared the same purpose and passion in life, we can accelerate the exchange of information and improve services for our clients."

As part of her work in China, she got involved with Sias International University, a private school started by Shawn Chen, a Chinese American entrepreneur with a vision of an East meets West campus that integrated Chinese and American teaching methods. Jerrie served as president of the Sias Foundation Board and initiated an Annual Women's Symposium attended by thousands of Chinese female students and their mothers.

She asked Shawn about starting a Women's Academy for the Future of Women at Sias. "I told him it would be a bold and daring leadership program to prepare women for campus, community, country, and global leadership. Shaw said, 'That would mean that the first woman president of China would be a member of the World Academy and have graduated from Sias.' I said yes, and he agreed to the program."

For Jerrie, action is the name of the game, and that is reflected in the curriculum at the World Academy. "If you create something and you don't take it to action, then nothing's going to come of it. I don't think women simply need more knowledge. They need to know how to utilize and activate what they already know to be important and powerful. When this occurs, the kindle of self-determination is sparked and the torch is lit."

What Extinguishes Her Fire

> *There is nothing that doesn't have resources, answers, or access to answers.*

Extinguish is not a word in Jerrie's vocabulary. She is, however, deeply disturbed by the lack of leadership focused on solving world problems. She believes we absolutely can solve hunger, educating children, providing maternal and child-care, and eradicating HIV/AIDS.

She adds, "When John F. Kennedy was president, we put a man on the moon. We had no spaceship, no fuel, no astronauts, no flight plan, and we achieved it in less than a decade. Leadership brought forth the resources to make it happen.

"The single ingredient to change is leadership, not money. We've given people money all my life; things have changed marginally, but the issues still exist and some have worsened. Give people permission to fulfill their dreams and give people access to others. Give them a leadership role to make things happen. That's what we do in the World Academy."

Why Firestarters Are Different

> *They have to have the capacity to be inspired—not to be inspiring but to BE inspired.*

Every day, Jerrie meets people she considers Firestarters. They are people who want to make their lives bigger and better, not only for themselves but also for other people. "They need to be curious about themselves and their circumstances. They have to be able to look out every morning and go, 'Oh my God! What's out there for me today? What can I do with this?' It's not a single moment. It's every moment that holds the possibility to make something happen. It isn't luck. It's leadership."

DAVID A. FIELDS

Consultant Guru Maximizes ROI and Value

If I can enable one hundred people and each of them helps one hundred people, then I'm helping thousands.

—David A. Fields[15]

David A. Fields has a passion for independent consultants. "Consulting is all about helping other people," he says. "My mission is to enable consultants to help others, to become a force multiplier—that way, my wins are multiplied exponentially through the work I do."

As 50 percent Innovator, 35 percent Initiator, and 15 percent Instigator, he is one of the world's experts on how to maximize return on investment (ROI) from consulting engagements. Author of *The Executive's Guide to Consultants* and *The Irresistible Consultant's Guide to Winning Clients*, he works with companies to get the most out of consultants and works with consultants to optimize the value they provide and receive from their expertise.

David grew up in Princeton, New Jersey, riding his bike past Albert Einstein's home. His father, an exceptional computer programmer whose solutions were elegant, creative, and logical, was a major influencer. Later in life, as David began developing the road maps and processes that would be at the core of his consultancy business, he would gravitate toward clarity and creativity in designing his methodology.

After obtaining his MBA from the Tepper School of Management at Carnegie Mellon, he brought his expertise in branding, sales, and market research to category management at GlaxoSmithKline. He left after nine years to become a consultant. "I went to a consulting firm that was led by a truly brilliant man who could sell ice to Eskimos. In the nine years I was there, I learned a lot about what to do and what not to do as a consultant."

His next move was to start his own consulting firm with a partner. David was the backroom engine, and his partner was the sales and relationship guy. After only four weeks, his partner left and David had a fledging business with no clients and with high expenses. His first year on his own was an "unmitigated disaster." Then a grizzled sales veteran from Philip Morris gave him advice: "David, you're actually a sales guy. You're very good at that. You can sell consulting."

He reinvented his company to sell not only what he did but also what other smart people did. "I went around meeting really interesting consultants.

Then I started talking to people I had met along my career and matched their problems to these smart people. That's how I built the Ascendant Consortium."

In addition to his corporate clients, he has also built a strong business helping hundreds of consulting firms worldwide win more projects from more clients at higher fees. He replaces the typical consultant's mindset of emphasizing expertise and differentiated processes with a focus on building relationships, engendering trust, and solving clients' existing problems.

"I love when I help another consultant or firm succeed," David says. "I get notes back from clients who share successes. 'You said to do this, and I did it and it worked.' Or, 'I landed this huge project or this huge client' or 'We're able to work with our employees better.' There's nothing better than putting yourself out there to help someone and having that person succeed."

What Extinguishes His Fire

> *It's after you get that first quick win and the second win doesn't come. Or after you get that second one and the third, fourth, and fifth don't come. Then what? That can knock a lot of people out of the game.*

David is fascinated by entrepreneurial drive and has spent years looking at what makes some consultants successful and others not. One area he is particularly interested in is emotional states. "There are a lot of the challenges in dealing emotionally with failure. Everything is easier if you're naturally optimistic. But what if you're not naturally that way? What if you see the worst in the world? Fortunately I'm fairly optimistic, but I've struggled with these issues my whole life. The thing that can extinguish the fire is lack of success because that is where your optimism, commitment, and resilience are tested."

Why Firestarters Are Different

> *You have to be willing to try something knowing you can fail. In fact, you have to be willing to try something knowing you WILL fail.*

David doesn't think intelligence and ability to scale are key differentiators for Firestarters. Instead, he uses words like drive, resilience, discipline, and fearlessness. "A lot of people have desire. A lot of people have dreams. The question is what separates those people who go past desire. Who will move past the dreams and actually create something?"

He believes that Firestarters who hit the right combination of place, time, person, and talent are rare. They are what Malcolm Gladwell calls the Outliers.[16] Instead, David suggests that we examine everyday Firestarters and see what makes them different.

"Your everyday Firestarter doesn't have that confluence of circumstance, event, and talent," David says. "It's more about commitment and courage. There's also a certain quantity element to success. You can't just try once, and you can't just do one thing. You have to try over and over and over again. You have to work at it and work at it and work at it."

DAME SHELLIE HUNT

Lifetime Achievement Awardee Transforms People to Design Their Success

> *Somebody reached out their hand to me [while I was] growing up in poverty. They eased me up and out. I always felt like it was an honor to be able to reach out to others.*
> —Dame Shellie Hunt[17]

Called the "First Lady of Entrepreneurs," Dame Shellie Hunt has been knighted along with Elton John, Bono, Bob Hope, Elizabeth Taylor, and Paul McCartney by the Order of St. John Russian Grand Priory, the oldest historical charitable order.

She is also a proud recipient of the Lifetime Achievement Award from President Obama; founder and CEO of Success Is by Design, ReMake MY Life, and the Women of Global Change; lead mentor in the Billionaire Adventure Club; a judge of the Gracie Awards; and board member for the National Women's Political Caucus.

As 70 percent Innovator, 20 percent Instigator, and 10 percent Initiator, she helps transform individuals and organizations by showing them how to design their success from the inside out. When Shellie was a young child, her mother built a bathtub out of used bags of concrete and leftover tile. The lesson she learned from the experience was not about lack of money.

"It was really the first time I watched entrepreneurial thinking because my mom had no instructions," Shellie says. "I asked her, 'Have you ever done this?'

and she answered, 'No.' But she could see it in her mind's eye. As a young child, I now understood that if you could see it, you could make it happen."

This led her to learn about human latent programming and the things that can stop or accelerate action. She created an approach to overcoming obstacles and reaching potential that has made her a leader among coaches and speakers. She has directly helped more than 100,000 individuals from all walks of life, ranging from CEOs and Supreme Court judges to business owners and inventors.

What Extinguishes Her Fire

The biggest challenge we have to face is ourselves. We get into the mechanics and start talking ourselves out of ever stepping into our passion.

Shellie believes it is easy to talk yourself out of something and let others influence you. She remembers as a young woman thinking about opening a clothing store. Her mother, who had never owned a business, gave her numerous reasons why it was a bad idea.

Later, as she became more confident in herself, Shellie became very firm about who she would listen to and why. "If I come up with [an] idea and one person starts to pooh-pooh it or act like it's going to be a lot of work, guess what? They can go find work somewhere else. I'm not talking about feedback. I'm talking about a negative attitude."

Because she defines herself as a passionate creator and visionary, she is very particular about who is on her team. "You become what you're around, no matter how hard you may try."

To prevent your fire from being extinguished, Shellie believes you need to pay attention to your inner voice. She stresses that the self-conscious mind opens up to deliver messages to the conscious mind like "I don't know how to do that so I can't do it." The trick is to have the conscious mind talk back to the subconscious.

"Go behind the curtain," she explains. "Do whatever you have to do to say to yourself, 'Oh, yes I can. Maybe I don't know, but I'll find somebody who does.' Start to bring it back to the subconscious mind, as the doors are open. If you start talking back to it, you're actually shoving the message back into the subconscious mind to reprogram it."

Why Firestarters Are Different

> *Passion is what lifts a car. When a mother lifts a car because a child is underneath, it is out of passion. That's where the magic and miracles happen.*

Passion is what makes Firestarters different, according to Shellie. "For me, it's a driving force, an energetic force. I often say to people, 'On one side of the room, there's a group of people sitting there solemn and quiet. On the other side, there's this group talking, interacting, and laughing. Which group would you rather hang out with?' People are attracted to light and passion."

When her own energy starts to wane, she fires up the music and starts to dance, which gets her teams laughing. Her dancing reflects her view of what drives a Firestarter. "Everything is energy and movement. Nothing is stagnant. Passion is energy in forward movement. One of the things with passion is you're going to look at opportunities differently than someone who lacks passion and only sees the obstacles."

RODNEY ADKINS

IBM's "$18 Billion Man" Fulfills Dreams through Focus, Discipline, and Diversity

> *There is a huge frontier of innovation. I always dream about these things.*
>
> —Rodney Adkins[18]

Rodney Adkins is a dreamer. He dreams about commercial space travel, teleportation, holographic communication, and regenerative medicine. Known as IBM'S "$18 Billion Man" and one of the people instrumental in the creation of Watson, Rodney spent thirty-three years at the company, leading global teams, managing multibillion-dollar business units, and delivering a vast portfolio of product innovations and enterprise solutions.

The first African American to attain that position of senior vice president at IBM, he is a trailblazer in innovation, engineering, and diversity. Since retiring in 2014, he has been an active board member at UPS, Avnet, PPL, and Grainger and is president of 3RAM Group, which specializes in capital investment, business consulting, and property management services.

His path to success was forged in childhood, growing up in a very tough part of Miami. Originally studying martial arts to protect himself, he discovered the added benefits of focus and discipline. He also discovered the power of daydreaming. "When I was growing up," he says, "I always put myself in very positive situations, and a lot of that was from daydreaming. I would think of places I wanted to be, things I wanted to do, or people I wanted to see."

He daydreamed about becoming an educator, politician, athlete, or entertainer but discovered that his special talent was critical and analytical thinking. By his senior year in high school, he knew he wanted to become a computer engineer and work for a company like IBM.

His road to success was filled with detractors. He remembers his first college advisor who recommended that he not major in science and engineering because his probability of success was very low. And there was his first manager at IBM who told him that his highest potential was as a technician. Being inducted into the National Academy of Engineering and reaching the highest ranks at IBM have been sweet vindication for him.

His refusal to accept the expected either in his own thinking or that of the teams he managed has been one of the foundations of his success. "Problem and need is addressed through diversity of thought where all the team members do not think the same, approach things the same, or have the same experiences in terms of their background. I look for diversity on every dimension."

60 percent Innovator, 30 percent Instigator, and 10 percent Initiator

> *I think the biggest thing that happened for me in my career was having access to talent and being in an environment where they allowed me to be Rod Adkins. I was a great employee but always edgy and never satisfied.*

While his work on personal computers, large enterprise systems, and supercomputers are well known, he feels his most important contribution has been in the area of mobility. "The concepts and techniques we worked on are the reason we have mobile solutions and operate in a connective environment with intelligent devices."

He believes the best solutions that came out of IBM under his leadership were the result of healthy intellectual debates done in a constructive manner. "I always wanted a team with debates, arguments, and polarized points of view. If you have polarization, it forces you to see things differently. If everyone on your team is homogenous and all agree, you may miss out on something significant or tremendous."

What Makes Him a Firestarter

> *I do have a pretty positive view of Rod Adkins. It may be distasteful to some or might sound arrogant, but, at the end of the day, you have to believe in yourself first.*

Rodney believes there are three core things that contributed to his success as a Firestarter. The first was that he believed in himself and what he could potentially accomplish. He would literally envision himself in certain situations and imagine how he could get it done.

Second, he had a strong support system that changed over time. "I have always had a strong support system throughout my life that helped me reinforce who I thought I was and what I could potentially do."

Third was fearlessness about taking on things that were not conventional or status quo. "I looked for opportunities where it was partially controversial, might be a breakthrough, or where others were unsuccessful. I learned very early that if you want to make it fast and big, you have to associate yourself with things that are hard to do."

Why Firestarters Are Different

> *Strongly trust your instincts and believe in yourself. If you really believe in your capabilities, more often than not you will be right.*

Rodney believes that Firestarters believe that nothing is a failure. "I will never forget things that we did at IBM that were complete failures, but they showed up later in ways that were much more practical. I have this view that nothing is really ever a failure; it was just not being conceived or implemented in the right application."

He also believes that Firestarters tend to pivot, not yield. "Nothing goes away with me. When it is clear that some passion I am chasing is not right, I apply the best aspects of what I was thinking to something else."

To do that, you have to accept that you will never be perfect, but always work at improving. "My batting average is not a thousand. It is probably .750. If it is .250, I keep going to get that batting average up. I am still in the game."

Chapter 28

INSTIGATORS DISRUPT THINGS

Without disruptions in life, where would we be?
—Sharon Goda, Canadian actress

W

e suspect that everyone has a little bit of rebel in them. Perhaps, it is a part of human nature. But Instigators are different. They are rebels on steroids. They live to change things. They often work in teams, but at their essence they see themselves as agents of change. They glory in it. As one of our interviewees put it, "I am a mapmaker. I draw my own path. I don't want a prescribed set of next steps."[1]

Of the people we interviewed, the largest number of Firestarters self-selected themselves as Instigators. We were not surprised. Being an Instigator is alluring. They embody courage to many of us. Tom Brady winning the Super Bowl against all odds because he was absolutely dedicated to changing things. Susan B. Anthony fighting for a cause she knew she would never see materialize in her lifetime.

Sometimes, they lead movements. They believe strongly in the words of former president Barack Obama: "Change will not come if we wait for some other person or some other time. We are the ones we've been waiting for. We are the change that we seek."[2]

Instigators also work on a smaller scale of change. They are the three-jobs-a-day moms going back to school to create a better life for their families and communities. They are the dying men and women raising awareness and money for a disease that will kill them before a cure. They are the entrepreneurs who haven't taken a paycheck in five years because they will not say yes to failure. They are the invisible men and women in the grocery store who crossed segregation lines over and over again. They are immigrants who build hospitals in their home villages and save the lives of children who will invent the next breakthrough water filtration system or new malaria medicine.

In this chapter, we share the stories of a range of Instigators. Some have enacted big changes that you know about; others will be a surprise:

- Noah Galloway: *Dancing with the Stars* Fan Favorite Instigates Change with "No Excuses"
- Dr. Marsha Firestone: Women Presidents' Organization Founder Helps Women Reach Farther. Together.
- David Egan: Special Olympics Global Messenger Advocates for People with Intellectual Disabilities
- Yasmine El Baggari: *Glamour* Woman of the Year Connects People One-on-One Worldwide
- Don Miguel Ruiz Jr.: Best-Selling Author and Teacher Finds Personal Freedom within Family Tradition
- Juliana Richards: Fashion Entrepreneur Shapes Her Dream into Global Success
- Scott Petinga: Cancer Thriver Delivers His Message: "F*ck Mediocrity"
- Dominique McGowan: Google Program Manager Fires up Teams
- Ziad K. Abdelnour: Economic Advisor to World Leaders Navigates Turbulent Times
- LaToyia Dennis: Motivated Mom and Education Advocate Lights up the Dark Places
- Patrick Ip: Digital Pioneer Asks, "How Can I Help?"
- Karen Benjamin and Joe Morone: Sales Instigators Show Nothing Happens until Something Is Sold
- Larry Boyer: Disruptive Technology Visionary Prepares for the Fourth Industrial Revolution
- Dr. Angela Marshall: Physician Heals Disparity in Women's Health
- Ezz Eldin El Nattar: Business Igniter Exposes Egyptian Businesses to the World
- John Salmons: Former NBA Basketball Player Defines Pivotpreneur

NOAH GALLOWAY

Dancing with the Stars Fan Favorite Instigates Change with "No Excuses"

> *People saw me working hard at something that was a struggle. This encouraged them to do more with their lives.*[3]

Few of us will forget the grace and courage of Noah Galloway's performance on *Dancing with the Stars* or the leadership he showed when his team was victorious on *American Grit*. A double amputee who lost his left arm and left leg in an explosives attack in the Iraq War, Noah has become a sought-after motivational speaker, coach, and author, with his message of "no excuses."

Turning weaknesses into strengths is a theme of his life that helped define him as 80 percent Instigator and 20 percent Initiator. A fitness fanatic and extreme sports competitor, he reacted to the loss of his limbs like many disabled veterans. He became withdrawn, out of shape, and depressed. Late one night, he took a long look in the mirror and realized there was more to him than the injuries. He committed himself to turning his life around. He got in shape and started running obstacle course races. This achievement earned him a spot on *Men's Health*, which led to *Ellen*, which led to his third-place win on *Dancing with the Stars*.

Raising three children also has been a key motivator of his "no excuses" philosophy. He tells a story about one of his sons who got all A's except for a C in history. "I told him that I wanted him to concentrate harder on history because it came harder for him," Noah says. "I explained that often people concentrate on their strengths, but the real success comes from targeting weaknesses and building off of those. I see it in the fitness world all the time where people overwork the front of their body and neglect their back because [the front is] stronger and [they] concentrate on those strengths. My son, by the way, got all A's including history."

Raised by a father who lost his own left arm before Noah was born, he remembers how incredible it was to see his dad work so skillfully. He also recalls that his father had weaknesses when it came to running his construction business: not asking for help, micromanaging, and giving too much away. "Now that I have my own company, I'm not afraid to step back and say, 'Look, I hired this person because they are good at something I'm not.' I'm not going to micromanage because then why did I even hire them?"

While his triumph over the explosives attack was the pivotal event that shaped him as a Firestarter, there were many other events that helped make Noah the man he is today. He remembers having his father ask his opinion about a room addition he was building. "To this day, I don't remember what I said, but the fact he asked my opinion put my mind in a state of 'Okay, what's the problem? How can we solve it? What's our objective?' So at fifteen years old, I was thinking in business terms."

Today his mission is simple: to help as many people as he can. "Whether a veteran, someone with a disability, or an able-bodied person, I'm constantly proving that they are capable of more than they automatically assume. Being in the public eye enables me to have more impact. I want to take it to the next level. I feel like the more I can build, the more people I can help."

What Extinguishes His Fire

> *I can come off very confident, but I have this side of me that*
> *doubts myself.*

People who see the confidence Noah exudes in public might be surprised that his Extinguisher is self-doubt. "It's something I've dealt with my entire life, and it still tries to creep in. I try and push it down. I use those moments of self-doubt to experience them, overcome them, and try to share that with others."

When he went back to the gym after missing his leg and arm, he would go at two o'clock in the morning so no one could see him work out. This insecurity of his body for the first time in his life gave him empathy. "I never dealt with obesity. Even when I was in the worst shape of my life during my depression, my body didn't put on weight. The struggle of getting back in the gym with my injuries let me relate and connect with people."

His biggest motivator in overcoming self-doubt is his children. "My family is more important than my own self-doubt. If I'm trying to work toward a goal to try and support my family, then my family is going to override that. That natural instinct to overcome takes over."

Why Firestarters Are Different

> *Whatever it is in life, don't let the fears creep in. Before they get*
> *too strong, say no.*

Noah believes Firestarters are people who pay attention and learn from mistakes. This means overcoming fears and weaknesses. To do that, you have to find the one thing in life that is more important than you.

"My children will always override my fears. It's taking that instinct to take care of your young and become more powerful than you ever thought you would be. You hear the story of the woman in the car wreck being able to lift the car to save her children. For people who don't have children, there is something in your life that you can say, 'This is more important than my fears,' and that is what you need to hold onto. That is what will get you through day after day to take those risks."

DR. MARSHA FIRESTONE

Women Presidents' Organization Founder
Helps Women Reach Farther. Together.

My goal is to have women who had achieved success not only keep going but grow more.

—Dr. Marsha Firestone[4]

Dr. Marsha Firestone is a rule breaker. You won't necessarily get that at first glance. She's polished and petite, and her roots growing up in Mobile, Alabama, give her the demeanor of a genteel Southern lady.

Then she starts talking about women business owners, and her 45 percent Instigator, 35 percent Innovator, and 20 percent Initiator nature emerges. As founder of the Women Presidents' Organization (WPO), she has built the premier global membership organization for women presidents, CEOs, and managing directors of privately held multimillion-dollar companies.

Behind her actions is a deep passion to bring economic security to women. "Women need to have economic security for themselves, their families, and their employees," Marsha says. "They need it because when you have economic security a lot of the social problems that are faced by women are reduced."

Twenty years ago she realized that economic security was directly tied to business growth. No organization was answering the vital question: what do women business owners need to grow? Today the two-thousand-plus women who belong to WPO in 137 chapters around the world have aggregate revenue of $24 billion with more than 153,000 employees.[5]

Marsha grew up during a time when there were very few opportunities for women and with a mother who had traditional views on gender roles. Her entrepreneur father was a major influence in her life, particularly when it came to education. His encouragement led her to earn a PhD in communication from Columbia University and become an expert in adult learning theory, nonverbal communication, and managerial competency.

Getting her doctorate is a clear example of the determination that Marsha has exhibited throughout her lifetime. One of history's great cultural anthropologists, Margaret Mead, was on the faculty of Columbia. Marsha sent her a letter asking for help on her dissertation and did not receive a response back. So while attending a lecture Dr. Mead was giving to three thousand people at the Museum of Natural History in New York, Marsha waited until they all filed out and then reminded her of the letter. Dr. Mead became her advisor.

A major career disappointment led her to form WPO. She had applied to be president of the American Women's Economic Development Corporation (AWED). During her interview with the board, she told them that if she got the job she would start an organization for million-dollar-plus women-owned businesses. "They did not give me the job, and I was devastated. One of my mentors was Bea Fitzpatrick who founded AWED. I told her that I really wanted to do it. She told me to stop complaining and go do it on my own. So I did."

When Marsha started WPO, she set unprecedented conditions for membership—$2 million in gross annual sales for a product-based business or $1 million for service-based. "No one had ever required these criteria. I broke the rules. The unwritten rules for women were that they shouldn't be worried about money. In fact, they shouldn't even talk about money. I didn't believe that was true."

The fuel that has kept her going is seeing the success of the women in her organization. "Nothing fuels my fire more than seeing women who are in their own businesses be really successful in accelerating the growth of their companies."

Also very significant is the support she sees members giving each other. "There was a member in one of our New York chapters who had breast cancer. The chapter took turns staying with her while she was dying. It was quite an unbelievable and unforgettable thing that they did."

Marsha's Accelerant is simple: her mission. After a career working to promote women-owned and women-led businesses, she still is appalled—and angry—by public perception: "The media does not believe there are women who own companies that are really substantial in size. They still think of them as mom-and-pop operations. I was watching the news recently, and they were bringing in people to talk about their businesses. Do you think there was one woman that they brought in?"

To change that view, she partnered with American Express for an annual ranking of the 50 Fastest-Growing Women-Owned/Led Companies. "I believe we are at a tipping point. We are proving to the media and population at large that women can and do drive very substantial businesses. They hire a lot of people and generate a lot of revenue and taxes. I want that recognition for women-owned and women-led companies."

What Extinguishes Her Fire

What stops you from being a Firestarter is when you lose your confidence.

Marsha believes that confidence and support are what enables Firestarters to keep going. That, in essence, is why WPO exists. For her, this support can come from many places—WPO members, sponsors, and staff and board.

One of her biggest supporters is her husband. "My husband is my greatest PR agent, and he's been very, very supportive. I really appreciate that." She also counts on a strong group of women friends. "I have known a lot of women in the business community for more than twenty-five years. We talk, share ideas, and learn from each other."

Why Firestarters Are Different

> You have to be able to survive those difficult days and keep on going.

Marsha has literally met and talked with thousands of women and men who are Firestarters. Her passion, career, and research have been deeply entrenched with understanding and helping people who ignite their lives and the world around them. At the end, she believes the difference between Firestarters and other people can be summed up in two words: emotional stability. "I think emotional stability is more important than being well educated, smart, or accomplished. Every day can be difficult, and you need a certain level of stability to keep going. That is the very big difference between those people who make it and those people who don't."

DAVID EGAN

Special Olympics Global Messenger Advocates for People with Intellectual Disabilities

> I hope that the next generation of people with intellectual disabilities will have a brighter future and great potential to be fully accepted and successful in their lives.
>
> —David Egan[6]

David Egan enjoys a life filled with family, friends, sports, a competitive job, hobbies, and travel. By the way, he also has Down syndrome and is a Star Trek fan.

His desire to live a rich life without societally imposed limits on what he can accomplish has led him to become a passionate and successful advocate. As the first person with an intellectual disability to be awarded a Joseph P. Kennedy Jr. Public Policy Fellowship, he made history by working on Capitol Hill with the Ways and Means Social Security Subcommittee. The 50 percent Instigator, 40 percent Innovator, and 10 percent Initiator is also a Special Olympics Sargent Shriver International Global Messenger, a National Down Syndrome Society ambassador, a board member to several disability organizations, and a public speaker.

Born in Madison, Wisconsin, while his parents were graduate students at the University of Wisconsin, his first years were challenging, but the university community was very supportive and rich with services. It was harder when the family moved to Virginia in 1984; inclusion was not a common practice in Fairfax County Public Schools.

However, at Vienna Elementary School and later in high school, he had many wonderful teachers. At home, his family treated him the same as their other three younger children. "I was taught that work is part of life," David says. "I helped with family chores and was not excused because of my disability. On the contrary, I engaged in all of the activities—the fun ones and not so fun."

He competed with the Vienna Woods swim team in his neighborhood pool and later in the Special Olympics. In 1996, when in high school, David was offered an internship at Booz Allen Hamilton, which turned into a staff position. "My first supervisor was great. She took it upon herself to teach me everything there was to know about being a clerk in the distribution center. She believed in me."

Twenty years later he continued to be a valued employee and was included in all aspects of the job from training and assessments to corporate events and meetings. CBRE became his new employer with Booz Allen Hamilton as his client when they outsourced their facilities and logistics management. With both employers, David received competitive wages and benefits. His supervisors supported his advocacy, which allowed David to volunteer his time, pursue his dream, and respond to speaking requests. In September 2017, David started a new job as community relations specialist with the Government Affairs team at SourceAmerica, a national nonprofit dedicated to finding smart business solutions for corporate and nonprofit customers in employing people with disabilities. This position fits David's advocacy, public policy experience, and desire to help people with disabilities find meaningful employment.

David is clear about what has shaped him. "My family, teachers, coaches, and mentors played a big role in who I am. People believed in me. They had expectations. Special Olympics built my confidence and inspired me to become a global messenger at the local, national, and international levels. Volunteering and public speaking gave me a platform to share my message of human dignity, hope, and inclusion."

David's mission is advancing the cause of people with disabilities. "It fits into the human and civil rights movement. I would say I am a trailblazer, a pioneer opening doors for the next generation of people like me. We are American citizens. We have rights. We matter. We can reach full potential if given an opportunity and people take the time to listen [to] and respect us."

He sees himself as a living reminder of a cause and a movement. "I work. I pay taxes. I am independent and interdependent. I am self-sufficient and, at the same time, I rely on my family and friends as they also rely on me."

He wants to follow the vision of Eunice Kennedy Shriver, the founder of the Special Olympics. "Her words and call for justice resonate in my heart: 'The right to play in any playing field, you have earned it. The right to have a job, you have earned it. The right to be anyone's neighbor, you have earned it.' She created a global movement that is changing our country and the world's actions and perceptions of people like me. My hope is that my efforts will extend her vision for human rights and her bipartisan call for compassion and dignity for all."

What Extinguishes His Fire

> *This is not only my mission. It is a task that takes many people to make it happen. I need to be working with many others to make it a reality.*

For David, not having the right connections or opportunities to achieve his mission are factors that extinguish his fire. He knows from experience that opportunities can put him in places where he can make a huge impact. "I volunteered at the World Games in North Carolina in 1999. I was assigned to help the media crew. That opened a great opportunity to have them discover me as an athlete and a leader. It was a pivotal moment in my life when they interviewed me and I was on TV."

That moment led Special Olympics leaders to notice him and began his journey of advocacy, travel, and public speaking for the Special Olympics. Other groups in the disability community were also impressed with David's advocacy and acknowledged his leadership with various awards. Later the Special Olympics invited him to join their first Congress Synode in the Hague in the Netherlands where he gave a speech that was simultaneously translated in five languages. That started his international outlook and travel.

Why Firestarters Are Different

David turns to another Firestarter he deeply admires to answer the question about what makes Firestarters different. "Gandhi said, 'Be the change you want to see in the world.' I am trying to live this, but I need others to help me." Instead of waiting for others to pave the path, David has been a trailblazer for others with intellectual disabilities.

YASMINE EL BAGGARI

Glamour Woman of the Year Connects People One-on-One Worldwide

> *I was crossing the Golden Gate Bridge, and it was such a symbolic moment in my life. No money, culture, resources, or social expectations would stop or hinder me from achieving my potential to have an impact on a global level.*
> —Yasmine El Baggari[7]

It is impossible to talk with Yasmine El Baggari without smiling. Her passion for her work is contagious, and you find yourself wanting to pack a bag and travel wherever she wants to go.

The Moroccan native, who is 70percent Instigator, 20 percent Initiator, and 10 percent Innovator, is focused on bridging cultures to encourage a more peaceful and caring world. That passion led her to create Voyaj. Voyaj.com connects people one-on-one from around the world to deepen global understanding and increase our empathy.

"My mission is to create spaces for people to understand each other," Yasmine explains. "Our experiences, education, and access to information divide us. Through conversation and authentic moments, we can bring our humanity back."

Along the way, she has been named *Glamour* Woman of the Year; been named one of the 100 Most Influential Travel Bloggers; made *Forbes*'s list of most inspiring and promising entrepreneurs; worked with the World Bank; done research at Harvard University; been an Engage America ambassador with the US State Department; spoken at the World Economic Forum and Global Entrepreneurship Summit; had her work appear in *National Geo-*

graphic, Forbes, and the *Huffington Post*; and received numerous entrepreneur-ship and ingenuity awards.

Her inspiration to travel came from watching Hollywood movies and from a strong connection to her uncle. "One day a French guy walked into my uncle's rural village and gave him a transistor radio, which he listened to every day. When he heard that a man had landed on the moon, he was awestruck and asked himself, 'How can a man land on the moon when I'm in the middle of nowhere doing nothing with my life?' That sparked his journey to become one of the most successful engineers at Monster.com. When he told us this story, I knew I wanted to come to the United States. I didn't want to follow a traditional path in Morocco."

Accepted into Hampshire College, she majored in economics and traveled to forty countries studying global migrations and other subjects. On breaks and weekends, she fulfilled a childhood dream to visit all fifty states, staying with friends and people she met on her travels.

Prior to her graduation, someone donated a million dollars to the college to encourage entrepreneurial endeavors. Yasmine came up with the idea of automating what she had learned from her travels into a system that matched people by their commonalities and values so that they could have conversations anywhere around the world with amazing people. Her pitch received $60,000 from the fund, and she then raised another half million to start Voyaj.

She is intent on using emerging business models as she builds out her company. "How do we create a structure where everyone wins from customers to stakeholders to team members? That's really an interesting inquiry for me. We have created a non-transactional model between the people getting connected around the world, but everything else is funded by third parties and corporations like airlines."

What Extinguishes Her Fire

> *I've been overcoming one barrier after the other since I discovered my fire. I'm a woman. I'm from the Arab world. I want to succeed.*

Yasmine does not see learning a new language, lack of money, gender, and geographic location as barriers, but more as opportunities to overcome challenges. "I realize this is a fun game of life, and we have the solutions within us. I stay true to my long-term mission and apply hard work, intuition, perseverance, emotional intelligence, and critical thinking."

But she does have fears and has learned how to overcome them from child development and psychology experts. "When I am experiencing fears or doubts, I trace it back to the past. This allows me to surrender and accept. I voice it, and it's gone. I can then go back to the truth. I don't pretend that I know everything, but what I've gathered so far allows me to be in a fire inflow state at all times, to pay attention to the triggers and fears, and not let them overwhelm and take over my life. It can be daunting, overwhelming, and hard, but it's true authentic power."

Why Firestarters Are Different

> *How do we spark self-awareness and inspire people to live aligned missions and lives?*

Yasmine believes that Firestarters are self-aware; they realize that there is more to life than their current reality. This awareness comes from being inspired by others in their community or being out of their comfort zone.

"I feel it's usually a breakthrough that happened for people at a younger age or midlife. They think success is XYZ. They achieve that success and then realize that they're not fulfilled with their life. That's when they switch 360 degrees and follow their path. Some people never get there."

For her, it was watching American movies and wondering if what she was seeing was real. For her uncle, it was a radio that was his access point to information that allowed him to dream bigger. Now with Voyaj she hopes to create that spark by exposing people in authentic ways to different environments and thinking in the world.

DON MIGUEL RUIZ JR.

Best-Selling Author and Teacher Finds Personal Freedom within Family Tradition

> *All of my family's teachings became relevant to my life. I stopped seeing it as something that belonged in a textbook or in history. I didn't have to repeat the way my grandmother or my father did it.*
>
> —Don Miguel Ruiz Jr.[8]

Don Miguel Ruiz Jr.'s journey has been one of rebellion against the expectations and pressures of being the son of a revered public figure, don Miguel Ruiz, as well as his search for his own voice within the family tradition.

One moment you will see the former Goth singing and dancing to Depeche Mode, and the next moment he will be talking about love, death, life, and knowledge. He is a Firestarter who defines himself as 50 percent Instigator, 40 percent Innovator, and 10 percent Initiator, and his books, lectures, and workshops have been life changing for a generation seeking personal freedom and optimal physical and spiritual health.

At the age of fourteen, don Miguel Jr. apprenticed to his father and his grandmother, Madre Sarita, who came from a long line of Toltec shamans dating back to the Aztecs. His father wrote one of the most popular and influential "self-help" books of the twentieth century, *The Four Agreements*. Millions of people globally, ranging from Oprah to New England Patriots quarterback Tom Brady, have followed don Miguel Ruiz's philosophy of how to live in the world through unconditional love based on Toltec tradition.[9]

Don Miguel Jr.'s childhood was defined as always seeing life from the diversity of different points of view. He spent much of his time crossing borders, growing up as a Mexican-American in Tijuana and in San Diego; the contrast of the two cultures shaped him as an individual.

His grandmother was a legendary healer drawing on spirituality and homeopathic remedies; his father and uncles were physicians trained in science and Western medicine. His home, where spirits, magic, and energy coexisted with logic, hypotheses, and experimentation, was built on love from family, relationships, and community.

Learning was also fun and playful. His father once told him, "If you want to learn how to play *Super Mario Brothers*, all you need to know is how to play with Mario. Learn to control Mario, because all the levels change and the one constant is always Mario. So is your life, Miguel."

Eventually the fanaticism and lack of privacy that surrounded his grandmother and father took a toll. The mother of one girl he dated as a teenager asked him to put his hands on her head and heal her headache. Another girlfriend was annoyed that his friends treated him like an average guy instead of someone special. As soon as *The Four Agreements* was released, people would ask don Miguel Jr. when his own book was coming out.

He rebelled against the family tradition and turned to his passion for music and movies. As part of film crews in Los Angeles, he earned a reputation as an energetic worker and moved constantly between gigs. He stopped working in film when he married his wife, Susan, and started a family. He realized that while he loved the high demand of movie business, being a husband and father were more important to him. The pivotal moment came when he asked, "What do I want my thirties to look like?"

He turned back to the family tradition with confidence that he could engage in his own unique way. *The Five Levels of Attachment*, for example, was a challenge to some of his father's fans because of its heavy use of sports analogies, something not generally associated with spirituality.

Today passion continues to drive him as he has ignited, fueled, and accelerated his dream. "As I finished writing one book," he says, "I am already working on the next because I have an idea. Boom, there it is. The passion I have in creating something is contagious. People hear that enthusiasm, and they love it."

What Extinguishes His Fire

Success takes on so many different forms, but it only comes with our effort to create it.

Don Miguel Jr. believes you always need to ask—what are you doing this for? "If the mission is great, then it's all worth it. A strong bottom line is important, but the mission is what motivates us to engage productivity with our passion and effort."

He puts this belief into action. When he goes to an event, he does not personally sell his books. "Why should I bring my own stuff to compete with the people who are hosting me? I want them to benefit. I want them to create. It's a team. They gave me the opportunity to talk to the people in their community. I am grateful for that, and I'm going to do the best that I can for them. We both win when we give each other the opportunity to do so."

Why Firestarters Are Different

I realize how short life is. Am I going to spend it creating something? Or am I going to spend it doing something I don't want to do?

For don Miguel Ruiz Jr., being a Firestarter happens when you are able to shed the image of how others see you and how you see yourself. "I became free when I realized that I didn't have to write a book. I didn't have to give lectures. I didn't have to live up to my grandmother or my father's image. I'm doing this because I'm enjoying it, and I want to do it. That changes everything for me."

JULIANA RICHARDS

Fashion Entrepreneur Shapes Her Dream into Global Success

> *When I think of my main purpose in life, it is to develop two incredibly young men who will eventually become fathers or whatever they choose to be. They will be very good citizens and contribute to this world.*
>
> —Juliana Richards[10]

Juliana understands not only what shapes the body but also what shapes the heart, mind, and soul. "I grew up surrounded by unrest," she says. "On the streets of Lagos, Nigeria, people were rioting in the midst of a political war. I remember hiding under the bed and hearing gunshots in the distance."

Today being a role model for her two sons is vital to the successful founder and CEO of BodyFab (formerly known as Slim Girl Shapewear), a fast-growing global brand that helps women shape their bodies naturally and includes swimsuits, waist trainers, and athletic wear.

In ten short years, the woman who describes herself as 50 percent Instigator, 25 percent Innovator, and 25 percent Initiator has created a body-shaping and contouring clothing line that has become go-to wardrobe necessities for high-profile celebrities such as Khloe Kardashian, Nicole "Snooki" Polizzi, Carmen Electra, and more.

Juliana arrived in the United States from Nigeria at age eighteen with one hundred dollars in her pocket, two sets of clothes, and two pairs of shoes. She learned about American culture by watching Jenny Jones, Ricki Lake, and Jerry Springer and developed a passion for her adopted country because it was a place where you could be whoever you wanted to be.

She has a deep confidence in herself instilled by her father who told his nine children to embrace wholeheartedly whatever they wanted to pursue. Her mother, who always had little businesses going on the side, instilled in her an entrepreneurial spirit.

BodyFab grew out of a need Juliana personally experienced because of pregnancy—how to naturally reduce belly fat after birth. She became obsessed with finding a solution and drew her inspiration from surgical clothes that were prescribed after a C-section and traditional body wrappers used in other countries to help women get their shape back after pregnancy.

One of Juliana's favorite questions is, why is it like this? Coming from a traditional culture and upbringing, she works hard to instigate change without being disruptive. "My culture is very masculine and feminine. Women always have their own place, and men have their own place. I will come into a room and acknowledge that things have been done in a certain way for years and years, but ask if we can change it a little bit."

Her desire to make change as a woman is epitomized in her venture into e-commerce in Africa. "In Nigeria to this day, I still do not hear word of a strong e-commerce platform that is headed by a woman. As we speak, I'm building a marketplace that sells women-only items. Because I have been successful with running my own e-commerce platform and selling just our brand, we're branching out to create a marketplace where other women business owners or men can sell items that are geared toward women."

Fear and competition are two factors that have ignited, fueled, and accelerated Juliana as a Firestarter. "I was in my late twenties and afraid that I wasn't going to accomplish what I really wanted. That fear made me stay up every night to study and learn about how e-commerce works. Even though we have had lots of success, fear still keeps me researching and revamping. I use fear to motivate myself."

She also strikes a balance about how she approaches her competition. "There are times I try not to look at what my competitors are doing because I don't want to do the same thing. But other times I focus on what areas they are growing in. Because I'm a super competitive person, seeing what others are doing inspires me to add new features and products that attract customers."

What Extinguishes Her Fire

> *I strongly believe the people you hang around will determine where you end up in life.*

Having the wrong people in her life can extinguish Juliana's fire. "I'm huge on the company that I keep. My dad always said that one bad egg corrupts good character."

Her husband and business partner is proof of how carefully she surrounds herself with the right people. "I strongly believe in not just finding love but finding love with the person whose goal aligns with yours and has the same dreams."

She has also learned how to break off from relationships that are not taking her where she wants to go. "It may sound harsh, but you have to be ready for breakups. To do this I routinely take inventory. Is this where I want to be? Is this how I want to live my life? Do I want this amount of money? Am I happy?"

She does this evaluation to make sure she and the company are on the right path and then adjusts her team accordingly. "I always make sure that each person aligns with the dream that we have. If they don't, I do not have a hard time cutting off that relationship and getting right back on track."

Why Firestarters Are Different

If you are aware that you can bring something to the table, then you can help someone else. You will do everything in your power to share it.

Juliana believes that the people who have not ignited their Firestarter potential do not believe that they have something to bring to the table. Being told by her father that she could make a difference gave her the wings to explore and then share her vision with the world.

"There are people who really believe they have something to share, something that can impact someone's life. When they believe, you can see them making an effort not just being quiet and fading away."

SCOTT PETINGA

Cancer Thriver Delivers His Message: "F*ck Mediocrity"

I hate bureaucratic bullshit and am compelled to revolutionize the world around me.

—Scott Petinga[11]

Scott Petinga values his mission above everything else. It may have started when he told a former boss that he had cancer and the response was, "Is this going to affect your work?" At that moment he decided he would always lead others with his humanity. This led to his resignation several months later after completing treatment and his eventual journey as a serialpreneur (one who runs multiple businesses) and socialpreneur (one who focuses on aiding society).

Categorizing himself as 50 percent Instigator, 40 percent Initiator, and 10 percent Innovator, Scott has gone on to launch a dozen different ventures; many are socially minded and designed to improve humankind. He believes business—the most powerful human-made force on the planet—must create value for society, not just shareholders.

One of his most successful ventures is a data-driven communications agency, AKQURACY, which differentiates itself from other ad agencies with a team that includes a meteorologist, a scientist, former military intelligence personnel, and a statistician. Within five years of its inception, AKQURACY earned a spot on *Inc.* magazine's prestigious List of Fastest-Growing Private Companies, and Scott became a semifinalist for *Entrepreneur* magazine's Entrepreneur of the Year Award. Not bad for a guy who graduated from high school with a D average.

His multiple nonprofits make an impact in diverse arenas. TH!NK DIF-FERENT is an idea factory designed to spark innovation in underserved areas of housing and healthcare. Fairy Foundation forges lasting memories for adults with life-threatening medical conditions. The Center for Advocacy for Cancer of the Testes International (CACTI) is an international advocacy network whose mission is to advance the practice, research, and education of testicular cancer.

In 2017, Scott authored *No One Ever Drowned in Sweat: G.R.I.T.—The Stuff of Leaders and Champions*, which gives advice on how to handle life's hurdles. In the book, he breaks down each trait of G.R.I.T.—guts, resilience, initiative, and tenacity—and how to successfully use them to reach goals.

The book shares his experiences and the wisdom of over fifty notable CEOs, entrepreneurs, nonprofit heads, thought leaders, athletes, everyday heroes, academics, and forward thinkers from all walks of life. He published under the banner of his own company after walking away from a book deal with a traditional publisher because he wanted more creative direction over his own story and had a vision of how to incorporate the stories of other renegade professionals.

Always the Instigator, Scott aims to prove naysayers in his life wrong. "Countless people have told me that I couldn't successfully complete a specific task because I was not educated," he says. "Or I would miserably fail because I hadn't done it before. The desire to prove people wrong drove me to achieve the unachievable."

He is also driven to save himself and others. "While I was deemed 'cancer-free,' the lack of empirical data and [the] overtreatment to cure me changed me forever. I became infertile for over a half decade, my hormones vanished, brain functionality diminished, and now I'm experiencing muscle atrophy." To that end, he is putting his money and resources to spur radical ideas in healthcare as well as getting vital treatment into the hands of those whose very lives depend on faster-acting and less expensive pharmaceuticals and medical devices.

What Extinguishes His Fire

Some people think I push the bar too far. However, they just need to get out of my way.

Scott's in-your-face attitude is his strength, and his use of profanity is an intrinsic part of his personality. In fact, the tagline for the Scott Petinga Group is "F*ck Mediocrity."

Sometimes, his approach can hinder success, but he doesn't care. "It is who I am. I'm very outspoken, and I don't give a f*ck what anybody thinks. I show up in jeans and a T-shirt no matter whom I am meeting with. I've been through way too much shit in life and don't need to pretend to be someone I'm not by putting on a suit and tie. I'm also going to speak my mind. That turns a lot of people off."

Why Firestarters Are Different

Someone must take a stand, tell the untold tales, and finally right the wrong.

Scott believes it is a Firestarter's responsibility to give back in ways that impact as many people as possible. "I call myself a superhero for a reason. I discover what is good in the world and try to change what is not. When I first became an entrepreneur, no one was there to guide and mentor me. So now I help others navigate the mountainous obstacles in front of them, sort of like a Sherpa."

DOMINIQUE MCGOWAN

Google Program Manager Fires up Teams

Cultivate curiosity and question your approach. Treat others with respect because the team is everything. Think about why you are doing something and focus on impact.
—Dominique McGowan[12]

At twenty-five, Dominique McGowan is a fast-rising achiever at Google with a deep passion for the power of people working together. At Google, she is involved with designing and deploying products at scale and managing the technical alignment of the company's biggest support teams.

Her career path working with nonprofits and start-ups and then joining Google reflects her 60 percent Instigator, 20 percent Innovator, and 20 percent Initiator makeup. She says, "My primary motivating factor is seeing something that has the potential to be different and being able to get people on board to make that happen."

She is the person in the room who challenges why something is being done a certain way, and she believes one of her greatest strengths is inspiring people to be better and persistent. "As you raise a team, start a company, or manage something with global impact, being able to do that on an organizational or institutional level is never easy. Having that grit and ability to motivate yourself and others to keep going is what I've seen as probably the most essential skill I've developed and recognized in others."

One of the Firestarter elements she relies upon most to fuel herself is connections. She says an "army of people" have provided her with help and believed in her abilities, and much of it was unexpected. Google has been one of the prime sources.

"Google has a fantastic culture around this. The first thing they tell you when you start is that every single person is only a coffee away, and it's absolutely true. Everyone is working 150 to 200 percent above what they were hired to do, and no one ever says, 'No, I won't help you.' I see that altruism in how the company operates, and I find it inspiring in how people operate."

She also has had to overcome personal challenges that have made her both highly empathetic to the needs of people and a strong leader of teams. "These are three incredible things you can tell someone when they are going through something tough, 'I am here. You are still you. It will be okay.' That is what I encourage people to do when they feel powerless in a situation and see someone struggling."

Her own challenge came in high school when she tried to get on the most competitive dance team in Southern California with only five years' dance experience. She finally succeeded on the fourth try. "It was validation that you have to fight for the impossible, and when someone says you cannot do this, you prove them wrong."

Unfortunately, achievement came with a high price not uncommon in the dance culture—anorexia. At Berkeley going to school full-time in her self-designed major in strategic philanthropy and dancing on a Division I dance team, her health began to take a toll. She realized she needed to take action. Through twenty to thirty hours of weekly treatment and determination, she was able to achieve full recovery from the eating disorder.

"I remember walking to dinner with my friends. The sun was setting, and we were laughing. I realized I had made it. My face was no longer cheek to cheek with the concrete. It was over, and I was finally moving on."

What Extinguishes Her Fire

> *It's very normal to question one's capability, but I think if there is any extinguisher, it's that. I don't think I've let that win ever, but it takes a lot of effort to talk back. I want to get to the point where I don't even hear it anymore.*

Because of her age, one of the things that can extinguish her fire is her belief in her own capability. She has rapidly overcome that both by personal work she has done and the supportive nature of her workplace. She shares the advice given to her by her manager at Google: "Don't question your capability. You either can do it or you can't do it. If you're here and giving it a shot, then you can do it. All you need to figure out is how much time you need and who else you should loop in."

Why Firestarters Are Different

> *We set unrealistic expectations with ourselves when we believe that everything has to go right the first time. It will not, and that is part of the adventure.*

Dominique believes that Firestarters are defined by the moment they are facedown on the concrete. "It's one thing when everything is going well, and someone hands you a great new opportunity. It's a very different experience when no one is offering you anything, and you are at a low point."

Firestarters know how to get themselves back up and keep going. They also know there are many moments when it will happen again. "You have to be comfortable in those moments, trusting yourself and building the ability to survive. You have to get back up and know that some of the most interesting stories are the ones that have not been told yet."

ZIAD K. ABDELNOUR

Economic Advisor to World Leaders Navigates Turbulent Times

> *We have been given a real chance to change the course of the*
> *nation from one of collapse and ruin to one of real hope, renewed*
> *prosperity, and regained honor.*
>
> —Ziad K. Abdelnour[13]

Ziad K. Abdelnour came to America as an immigrant from Lebanon with little money but a big dream. Today he is an economic advisor to the world's top political, investor, and corporate leaders as well as average Americans. He shares a strong, and often controversial, message of how to financially navigate what he describes as "the troubled geopolitical waters of the post-crisis world."

The credentials of this man who categorizes himself as 70 percent Instigator, 20 percent Initiator, and 10 percent Innovator are impressive. Wharton MBA in finance. Thirty years' experience backing companies with an aggregate worth of over $10 billion. Author of the best-selling *Economic Warfare: Secrets of Wealth Creation in the Age of Welfare Politics*. More than a quarter million Twitter followers. Honored as one of the Top 1,000 Most Connected People in the World and as a Global Shaper.

He uses his platform to instigate change through his writing, speeches, and social media outreach. He uses his influence to have other people with power open their minds to different ways of viewing the world. Now he is intent on educating all Americans about how to view the world differently and build a wealth mindset. A long-time supporter of President Donald Trump, he outlines these ideas in his new book, *Capitalism from the Inside Out: How the 99% can Build Wealth, Fight Crony Capitalism, and Restore America's Promise*.

His heritage and family were instrumental in shaping Ziad. "The Lebanese are seekers of the truth," he says. "They also rarely work for other people because they are very entrepreneurial and independent."

Ziad attended private school in Switzerland and was chauffeured around in a Rolls-Royce. But this was not the life he wanted. "I didn't want the easy life. I wanted to be a self-made man like my father. I wanted to do it on my own."

He believes Firestarters are relentless. Originally rejected from Wharton when he was twenty-two, he spent five hours convincing the admissions head to change his mind. The decision changed because they saw how persistent and strong he was.

Raising his own children well and providing guidance for future generations is vitally important to him. And this means having a strong America. "The abuse of power and the divisive politics has reached a tipping point. The abuses have been hidden for so long that if America continues on its current course, the empire of the United States will not be sustainable."

What Makes Him a Firestarter

Become powerful enough that the blowback doesn't scare you.

Ziad believes in many of the elements that ignite, fuel, and accelerate Firestarters. Prestige, power, and connections are three that rise to the top. When misused, these elements lead to what he calls "crony capitalism," a state that he believes is the downfall of America. The way to combat that is for new thinking from Firestarters who have the power and money to make change.

The Accelerant that keeps him active as a Firestarter is his mission to overcome the "deep state," which he defines as the part of the government that lies beneath the surface and never changes, regardless of which party controls the White House or Congress.

"I believe it is the first time in the modern era that the man that now sits in the Oval Office has been and continues to be an adversary of the deep state, rather than a tool of it. He is not a part of the establishment and has made obvious moves towards fighting it."

What Extinguishes His Fire

When you talk about the future, you are young. When you talk about the past, you are old. Tomorrow needs to be better than your yesterday.

Ziad believes strongly in what America stands for and what it has achieved in the past. But his focus is totally on the future. "We must have liberty and justice for all as one nation, powerful, indivisible, with opportunity for all to live the American Dream. More than ever, everyday Americans need to understand how to open their eyes in order to see the world differently."

His fire is extinguished when Americans adhere to a defeatist mentality and refuse to become informed. "It is time to turn off your social media, take your eyes off your smartphone, and start paying attention to what really matters. It's time the American people educate themselves on the state of our country and become part of the solution."

LATOYIA DENNIS

Motivated Mom and Education Advocate
Lights up the Dark Places

> *How can I be a light and help you understand that relationships are your most valuable resource?*
> —LaToyia Dennis[14]

LaToyia Dennis grew up with strong women influencers who ignited her desire to help mothers and children. "My grandmother was absolutely amazing," she says. "She just loved me unconditionally. Her hope for me fueled my faith because she believed I could do anything and be absolutely anything."

Additionally motivated by her husband and son, the woman who describes herself as 50 percent Instigator, 25 percent Initiator, and 25 percent Innovator created two ventures that have impacted the lives of thousands of people. A Chance to Learn is a nonprofit whose mission is to find and provide solutions to enhance families through women empowerment and quality education for children.

"A Chance to Learn was inspired by my son. He was two and wanted every child to learn how to read because if they didn't know how to read, then they couldn't read their Bible."

Motivated Mom, an online motivational, inspirational, and educational community, helps moms become great women and extraordinary moms. "If you're financially strapped, you can't think of anything other than how you're going to pay the bills and feed your child. So when you can live a life of purpose in abundance and financially free, you will be a more engaged parent. That's what the Motivated Mom platform is all about."

Growing up she wanted to be an actress, but as she got older she realized she wanted to make a difference in people's lives. "I wanted to be a catalyst for change. I wanted to be a person who enhanced the life of someone else."

LaToyia spent nearly twenty years as a nonprofit management and fundraising professional, raising millions of dollars to help further the mission of individuals and nonprofit organizations. Now she is slowly reducing her consulting practice and investing all her time and resources on A Chance to Learn and Motivated Mom.

"I know what I want to do for the rest of my life—to further early child-hood education and make programming available for kids so they are prepared for kindergarten and beyond. And to encourage moms to raise better children by being present and engaged in their child's education and life."

The opportunity to make a difference is what fuels LaToyia's fire. "I want my life to be impactful. I want my life to have mattered to someone other than myself and even my immediate family. I want to be impactful in changing how we educate three- to five-year-olds."

To accelerate this passion, she believes that collaboration is critical. To that end she is a strong proponent of collaboration among nonprofits. "If Motivated Mom partnered up with the Positive Mom and the Crafty Mom, then you have all of these people coming together with their ideas, thoughts, and passion. Now we are making one big giant impact on moms; the reach is further, and the knowledge is more in depth. When we are united, our world can be better."

What Extinguishes Her Fire

You just have to know where you come from and what your buttons and triggers are.

People in her life have questioned LaToyia's ability to accomplish what she wants. She believes this comes from their need to control her. "I find ways to overcome that manipulation and their desire to make me feel less than."

She believes there are people who could be Firestarters who feel defeated by the negative environments that surround them. "If you grow up in a home where you were not hugged or told you are loved, these things have an impact on you. They can determine how you will live and pursue your life's purpose. You know what you should do but feel so unimportant and insignificant that you don't chase after it."

Those who ignite the fire inside of them have heard a message and are encouraged. "They are told they can do anything and are smart. They are told 'I believe in you,' and it changes everything for them."

PATRICK IP

Digital Pioneer Asks, "How Can I Help?"

> *My goal every single day is how can I be a better giver.*
> —Patrick Ip[15]

Patrick Ip's mom and dad had a very clear idea of what they wanted for their son—a top-notch college education and stable high-paying job. By age twenty-six, the Firestarter who is 50 percent Instigator, 30 percent Initiator, and 20 percent Innovator has achieved their goals but with his own twist.

Currently, he is cofounder and CEO of Unity Influence, a new artificial intelligence technology company that matches microbrands and micro-influencers. He left a job at Google where he worked with thirteen Nobel laureates on the 1 Billion Acts of Peace initiative, which has been nominated seven times for the Nobel Peace Prize.

He presented at TEDxHongKong and is also founder of Ballers, an online and offline global community whose members focus on one question: how can I help you? While at the University of Chicago, he started and sold a social media firm that was named one of *Inc.* magazine's Coolest College Start-ups. And in high school, he was California State director for students for Barack Obama and a keynote speaker for a major United Nations conference in Australia.

Growing up in Modesto, California, he was the product of what he describes as "your stereotypical tiger-parent household." He remembers extra homework in kindergarten and being drilled on college admissions essay questions in third grade.

"We were in the car on the way to school, and my mother asked, 'Patrick, what do you see outside this window?' I said, 'Oh, people and trees and cars.' And my mom responded, 'No, that's the wrong answer. You're supposed to talk about innovation and your view for the future.' It's funny now because I work in the innovation space."

On his way to attend George Washington University, he stopped at the University of Chicago's admissions office and presented them with a long paper about why that school was perfect for him. He got in and went on to a stellar college career that included heavy involvement with the local start-up community.

Patrick loved working at Google. So many people were surprised when he left. He said a pivotal moment was talking to a friend who started Rip van Wafels in college and is now twenty-nine. His waffles are featured in Starbucks, grocery stores, and other establishments throughout the United States.

"Rip said he thought he only had five good years left. He wanted a family, and even though he loved the food industry, he didn't feel he could be at the top of his game with other major distractions and commitments."

This drove Patrick to leave Google and start another start-up while he still had the freedom of not worrying how to support others. He also was motivated by another factor—making the world better. "We assume every single year that the world is going to be a better place, but we always assume other people are going to make it happen. You have to be the one to create that change. And in many ways, going into start-ups is in my way of trying to shape the future."

What Extinguishes His Fire

> *You're putting everything you have into building, but when someone that you sincerely trust and love is telling you, "No, you should run the other way," it creates this huge conflict.*

Patrick's family is extremely close-knit and has sacrificed a great deal for him. "It's been so important for me to have their buy-in when I do endeavors, but I did not have their support when I was trying to do start-ups, and that was a heavy strain for me."

His parents wanted him to be a consultant or investment banker since those were considered prestigious and high-paying jobs. They were also highly concerned about stability. Patrick, on the other hand, was instinctively drawn to being a social entrepreneur and working in start-ups.

"It was really tough going through college and balancing this relationship with trusted advisors and my parents. I rejected a lot of the advice from my parents, which caused a lot of strain. I needed to move away from their perceived ambition for me and figure out what that ambition was for myself."

Today his relationship with his parents is built on trust and respect. He admires them and has empathy for why they wanted a different life for him. They have even been able to accept his decision to leave Google. "My parents have come to respect more of my own ambitions for my own life."

Why Firestarters Are Different

I don't think it is intelligence that separates us. I think it's truly hard work and the belief that you can do it.

Patrick believes that being a Firestarter is all about overcoming obstacles. Many people, however, ask a precursor question to "How can I get over the obstacles?" They ask, "Am I capable? Am I good enough?"

He has never asked questions about his competence or confidence. When he was growing up, he assumed he could play sports like soccer or baseball. "I didn't know anything about them, but I knew I would figure it out. I actually believe that you can go out and do lots of things. I believe you can have the confidence to own your future."

That means that the hardest question people have to ask is, do you believe you can do it? If they can answer yes, then they can be a Firestarter. "If you believe you can do it, then you can go straight forward to the second question: how do you do it? That is a much easier question to answer actually."

KAREN BENJAMIN AND JOE MORONE

Sales Instigators Show Nothing Happens Until Something is Sold

I need to be part of things that create freedom for others. When that is happening, my fire is lit.

—Karen Benjamin[16]

It was November 2007, and Karen Benjamin and Joe Morone finally had realized their dream. They had left behind highly lucrative and successful careers in executive-level sales leadership at Ciber, a global leader in IT solutions, to form their own new company, Worldleaders. With top-tier clients, enthusiastic employees, and strong revenue, they were on their way to fulfilling their ten-year vision of a business focused on new models for recruiting.

Enter the Great Recession. Recruiting? Try selling underwear to a nudist. Karen recalls it vividly: "Joe and I were sitting there looking at each other. It was probably one of the scariest moments of my life, but I knew that our business was going to be even better than it was before."

A powerful belief in each other and their combined drive to succeed are two of the key reasons why the life partners and cofounders have turned Worldleaders into a fast-growing sales consultancy, training and recruitment organization that helps CEOs and sales leaders of B2B technology companies.

Joe characterizes himself as 100 percent Instigator while Karen is 50 percent Instigator, 30 percent Initiator, and 20 percent Innovator. For Joe, being an Instigator is about how he approaches learning. "When I want to get good at something, I watch the best of the best and their competition. What I want to do is learn from both of them to recreate something new. That is how we wrote our book *The Smart Sales Method*. We looked at three of the best sales methodologies out there and compared them to our thousands of hours of research."[17]

For Karen, being an Instigator is about asking questions and taking actions. "If I do not like it, I cannot shut up," she says. "If it does not make sense to me, I will not stop. I think by asking questions. That has been a pattern my whole life, and there good and bad things have come from that. I am that person who can instigate and initiate the actions that make it happen."

They both see their mission in life as maximizing people's potential and believe working with CEOs is a great way to accomplish it. As Joe explains, "It is a desperate feeling in me to want to see you at your best potential. That is important to me for everyone I am dealing with."

For Karen, maximizing potential and her love of sales started at age thirteen when she created a babysitting service that profiled families and then matched babysitters to their needs and interests. "When I was looking at the people who were babysitting, many were doing a crappy job. I saw it as a mismatch between what the kids needed and what the babysitters were providing. I wanted to make connections that were the right fit."

For Joe, maximizing potential is rooted in wrestling. "I was an undersized kid in a tough school. Getting involved in high school wrestling enabled me to defend myself and develop a level of confidence. I became a northeastern national champion at age thirteen. It made me feel like someone. It gave me the ability to fear no man."

He gained his love of sales by watching his highly successful uncle, a sales manager at a Ford dealership. One day his uncle gave him a one-hundred-dollar bill, an ad for Greyhound, and this advice: "Joe, when you can sell, you can make it anywhere in the country. You get on a bus, off the bus, and walk down Main Street to a six-figure sales job. That is financial freedom, the freedom to live your life on your own terms."

Prior to joining Ciber, where he met and teamed up with Karen, he worked for a software development firm that was an IBM business partner. "CEO Tom Watson said, 'Nothing happens until someone sells something.'

Today at Worldleaders we wear wristbands that are modified to say, 'Nothing happens until something is sold.' It is all about teams and being able to bring in revenue for your company, self, and family."

Karen was the only business partner Joe ever wanted. The reason is because of the intensity and hard work she brings to the business every day. They agree that Firestarters in business together must have a great deal of trust. Karen recalls when their business almost failed: "I remember sitting in that room knowing there was no one else I wanted to be sitting there with. I knew we were going to find a way. Joe is a person who brings that out when you need the courage or push."

Why Firestarters Are Different

> It is absolute fantasy and total bullshit when we hear that the better people do not break or quit. I have been around the best, and I can tell you everyone breaks. It is how quickly you can recover that is the single most important piece.
>
> —Joe Morone

Joe has thought a lot about what makes people succeed. "I look at three simple things," he says. "First, the courage to pick the things you believe are going to work for you and throw the ones out that will not. Second, knowing that you will break, quickly fixing it, and getting back in the game. Finally, if you have done one and two correctly, you have earned the right to feel better than the people you compete against. Use that to stand up in competitive situations."

Karen agrees with Joe's analysis and puts specific emphasis on staying in the game: "I play when broken. Literally, when my arm was broken, I still came to work and kept going. I do not quit."

LARRY BOYER

Disruptive Technology Visionary Prepares for the Fourth Industrial Revolution

> My passion is helping people understand the implications of the fourth Industrial Revolution—all of the automation, robotics, and other new technologies that are coming—and how it's going to impact their lives for better and for worse.
>
> —Larry Boyer[18]

Larry Boyer grew up as a science geek in a rural community, a dedicated Boy Scout and one of only thirty-two students in his high school graduation class. "Everyone knew each other," he says. "My own interests were a little different from everybody else's. I had a very deep interest and passion for science. I was very curious about the world and how it worked."

In college, he majored in physics but became increasingly intrigued with economics, getting his master's degree at Rutgers and then earning another master's in public policy from George Washington. This background now provides the foundation for his mission to help people overcome their fears of disruptive technology and use it to their advantage.

"Fear comes from not understanding what is happening and what to do. If you feel there is nothing you can do, that amplifies the fear. If you get rid of fear and helplessness, you can relax and develop your strategy for moving forward and taking advantage of all the opportunities."

To that end, he has become a leading expert and global speaker on anticipating, preparing for, and responding to the volatile and dynamic changes caused by disruptive technologies, economic turbulence, and omnipresent business failures.

One of his core beliefs is that leaders need to recognize the right opportunities and act quickly. "There are times when you need to make a decision and go with it. It's important to know your own vision, mission, and purpose so you recognize an opportunity. It's easy to get sidetracked on something that is not right for you."

45 percent Instigator, 35 percent Innovator, and 20 percent Initiator

> *The machines and programs providing efficiency also threaten to radically change the lives of everyone who must now learn to keep pace or risk falling behind.*

Larry has developed a road map to help businesses thrive in an economy while others struggle to survive. He believes it starts with tapping into the entrepreneurial spirit that is driving innovation and disruption, capitalizing on the rising trend, and building what he calls the "Business of You."

"Globally, we are advancing headlong through the fourth Industrial Revolution—a revolution of analytics and technology—that consists of data-driven smart products, services, entertainment, and, yes, jobs. These advances open a wealth of exciting possibilities, but they also promote grave concerns."

He is particularly concerned about the number of jobs that will disappear due to automation, artificial intelligence, and robotics if businesses do not adapt. "A report from the World Bank says that 70 percent of jobs in the developing world could be lost. The US also may fall victim, losing up to 40 percent of its jobs. If so, we will not see the promising changes in society, government, and employment that so many desire. New jobs will be created, but if you are not ready for them, it doesn't matter."

What Makes Him a Firestarter

> *I'm not dependent on anybody's approval or assessment of what I'm doing because if I'm feeling strong and confident, I don't need anybody else telling me it's good.*

Larry's passion is his curiosity about things and, particularly, the connections between them. "I'm curious about understanding how the world works and want to share the connections I see between things. I see more connections than most other people."

His passion is tempered by pragmatism and resilience. "There are lots of things that are interesting, but if it doesn't really have an impact, why spend your time on it? And if something isn't working, I look someplace else. You have to be resilient."

Why Firestarters Are Different

> *We are the average in all ways of the people we spend the most time with. Part of success is making sure you are with people who can help elevate you.*

Larry believes there are two primary things that make Firestarters different. The first is a deep passion. The second is having the skills and people needed to help amplify your accomplishments. "It's very hard to do something all by yourself. It's hard to learn the skills to actually make something happen. You don't learn that in school. So who are the people who can help?"

DR. ANGELA MARSHALL

Physician Heals Disparity in Women's Health

> *I've always felt like there was something special about me. The circumstances of my birth made me want to do extraordinary things to prove my legitimacy. Those extraordinary things got me attention and opportunities.*
>
> —Dr. Angela Marshall[19]

Dr. Angela Marshall felt the sting of abject poverty throughout her childhood. Her greatest fear was being homeless—eviction notices were a regular part of life with her single mom. This harsh world formed the woman who defines herself as 50 percent Innovator, 30 percent Instigator, and 20 percent Initiator into a staunch advocate for economic independence, women's health, and self-care.

Today she is the founder of Comprehensive Women's Health (CWH), a thriving award-winning medical practice that has seen more than twenty-five thousand patients since its inception in 2007. "Women are vulnerable, because they often put themselves last on the list when it comes to self-care," Angela says. "This is amplified for women of color. When health fails, the risk of losing economic independence significantly increases."

As an African American female physician—a group that makes up only 2 percent of all doctors—she is a sought-after expert nationally due to her passion for these issues. A frequent contributor on CNN, *Fox5 News*, and *Let's Talk Live*, she shares advice on a variety of health issues that affect women differently than men such as depression, weight gain, heart attacks, and arthritis.

Angela is also a savvy businesswoman, marketer, and entrepreneur. To fulfill her mission to be a vibrant leader in women's health, she constantly explores ways to grow such as collaborating with other entrepreneurs, opening new offices, developing cutting-edge products and services, and writing a self-help book aimed at improving the health of women.

Her own way out of poverty was by excelling in school. From an early age, she dreamed of becoming a doctor but was reluctant to take the financial risk associated with the high cost of medical school. Instead, she earned a degree in electrical engineering and was awarded the prestigious CIA Stokes Scholarship, which covered all her undergraduate expenses.

She worked as an engineer for the CIA for two years before leaving the agency and going to med school. "Being affiliated with the CIA made me stand out on my medical school application. Becoming a doctor and starting my own practice let me help women and fulfill my passion for business, science, and math."

Angela's love of engineering makes her fascinated with solving problems. Her biggest innovation so far was born out of her desire to change the way women are treated by the healthcare system. "Creating CWH was very personal for me. Because I understand what women want and need, we were able to create and market authentically, which was key in helping us grow so quickly."

While treating patients, she is always challenging the status quo. Because she treats many victims of sexual violence, she developed a model for an emergency alert system that can easily be activated to reach close contacts and authorities to help thwart attacks. In another instance, after noticing a high incidence of nail infections, she created a prototype for a nail-cleansing system with a disinfecting process to help reduce the spread of fungus and bacteria in nail salons.

She remembers watching one of her heroes, Oprah, interview Mark Mathabane about his intensely inspirational autobiography, *Kaffir Boy*. "He said, 'Luck is being prepared for opportunity.' It was a statement that made such an impact on me. It makes me do my best to be prepared at all times for those lucky situations."

What Extinguishes Her Fire

You have to believe that there's always a way. You have to plow through.

Angela's past has been filled with moments of deep tragedy and remarkable fortitude. The most poignant was the senseless death of her four-month-old son when she was in medical school. Because the doctor on call in the emergency room refused to listen to her, her son died because he was not treated quickly enough. Within a month of his death, Angela had to complete her rotation at the very same hospital where he died.

"My son's death crushed me. It took a long time to regain my self-esteem, my confidence, my everything. I cannot even find the words about how difficult it was to go back to the same hospital where he died. But I had come so far. I knew I had to finish no matter how painful [it was] in order to graduate med school."

This experience reaffirmed her ability both personally and professionally to muscle through situations with willpower and perseverance. "It's a really important skill because we all have difficult times. Being able to keep grinding when your back is against the wall is something that is necessary."

Why Firestarters Are Different

> *I think it boils down to two things. One is hope, and the other is fear.*

Angela believes that lots of people operate under fear, but Firestarters operate with hope. "Even as a child, I had hope and the feeling that anything is possible." She believes that non-Firestarters often are naysayers.

"No one was offering to pay for my college. Yet, these same people tried to instill this fear as to why I shouldn't take a totally full-ride scholarship from the CIA with a guaranteed job after college. I didn't listen to their fears, and it was one of the best things that ever happened to me."

To this day, she never forgets where she came from and who has stood by her—family, friends, colleagues, patients, and God. "I still remember the little girl who never saw a doctor growing up because we had no health insurance. I still remember the acute pain of untreated migraines that were not diagnosed until I was in med school. I refuse to live in a world where I cannot do my best to alleviate the unnecessary suffering of women—mind, body, and spirit."

EZZ ELDIN EL NATTAR

Business Igniter Exposes Egyptian Businesses to the World

> *I want to become the person who has ignited an eternal flame in the lives of my children and as many Firestarters as possible.*
> —Ezz Eldin El Nattar[20]

Ezz Eldin El Nattar is a business development wizard with over twenty years' experience creating and managing start-ups from conceptualization to maturity. His primary stomping ground is his birthplace Egypt, which he describes as "the land of opportunity."

Understanding people and how the system works is critical, according to the CEO of Vantage Business Development Solutions and head of Ascendant Global, the global division of the Ascendant Group. "It is not enough to know people," Ezz explains. "You have to understand them and what they are capable of, where their downfalls are. At the end of the day, it is all about people, just people talking to people."

Ezz spent his first eleven years in Abu Dhabi in the United Arab Emirates where his father was chairman of a construction company. He was exposed to people from every country, working together to build the new country. Returning to Egypt required a great deal of adaptation to a very old country with deep history and traditions.

His multicultural experience was amplified when he received his MBA in international marketing at the University of Dallas. Since then the 50 percent Instigator, 25 percent Innovator, and 25 percent Initiator has worked for several corporations. As an entrepreneur, he has started and grown multiple businesses, ranging from building a trade portal for thousands of Egyptian exporters to creating a natural pet food brand.

"Most of the businesses I went into over the past seventeen years as a partner or consultant were new ideas and start-ups that always fascinate and trigger me. Anything that needs creativity and out-of-the-box thinking is my thing. It's what motivates me."

Ezz is fascinated with changing things for his clients and in the marketplace. "Changing things is something that requires a lot of finesse because people are not usually happy with change. It requires more than new ideas and thinking outside the box. It is about convincing people to change."

He contrasts that with being an Innovator. "When you start something new—an invention that no one has done like our Trade Egypt portal—it is a big challenge when you cannot compare it to other things. It's more about building the need and proving that this new thing is something people will benefit from."

He keeps the fire burning through progress and evolution. "Going from infancy to adulthood to maturity—just watching the thing grow—is always something that fuels my passion. As long as I am seeing progress, I am always giving it more."

He also advocates collaboration to accelerate success whether it is with his internal team or external partners. "I always believe there is a point where you have to join forces with other parties. You can't stay on the track by yourself forever."

Building a strong internal team from the beginning is vital. "If you can master the people part, you can master the world. Once your internal team is on track with you and shares your passion, you can sit back and be confident that your team will do exactly what you would do when you are there. This is how companies grow."

At the end of the day, he thinks the CEO is a primary difference between a good company and a great company. "Think of companies like NASCAR vehicles. They all have the best equipment and essentially are 95 percent the same. The real differentiation is the driver."

What Extinguishes His Fire

> *A lot of people would say I am too much of a perfectionist. It is something that makes collaboration trickier because I want to do something in its best form. It is not always the ideal case, but to me it is always the ideal case.*

Ezz is aware that he has high standards and can have his fire extinguished if others don't share his desire to be the best. "When I get that feeling that I am not getting moral support or my partners are not sharing my vision, I start getting demotivated. I am not going to end up doing something that is just another thing. If I go into an existing industry with competitors, I want to be the best in it."

Why Firestarters Are Different

> *You cannot really ignite someone's fire if they do not have the gasoline inside them.*

Ezz believes that Firestarters must have a fire inside them. He acknowledges that some people do not, and they are happy. "It is very important to understand that being a Firestarter is a desire and journey that not all people have the need to pursue. There are those who are just standing outside the fire. They are happy there."

He says some people are born with that fire, and others acquire it along the way. "If the fire is there, then you need to find out what triggers these people. Where is the gas leak that can ignite so it will just flow over their whole system?"

JOHN SALMONS

Former NBA Basketball Player Defines Pivotpreneur

> *Effort and working hard is what I'm used to. I had to constantly work my way through different barriers and obstacles to get to where I wanted to go.*
>
> —John Salmons[21]

John Salmons has shaped basketball, and basketball has shaped him. During his thirteen years in the NBA, John played guard and forward for the Philadelphia 76ers, Sacramento Kings, Chicago Bulls, Milwaukee Bucks, Toronto Raptors, and New Orleans Pelicans.

Known for versatility, reliability, and tenacity, he is now applying the qualities that made him a world-class professional athlete to businesses and social initiatives. He says, "I discovered the same approach on the court of being able to move with fluidity doing multiple things at once brought a great deal of value to how I saw business. I call it being a pivotpreneuer. I am focused on bringing my 3-D formula of desire, diversity, and discipline to the entrepreneurial world. My intention is to build an off-the-court empire that leverages my passions in technology, entertainment, fashion, real estate, publishing, and franchise ownership. Being able to make the world better from a Christian standpoint is my ultimate goal."

As a social engineer, he describes himself as 45 percent Instigator, 35 percent Initiator, and 20 percent Innovator. He is involved in Three Squared, a company that utilizes shipping containers to build commercial and residential properties; Fan Cheer, an interactive platform for televised sports entertainment where fans compete to demonstrate passion for their favorite teams; and UFan8, a live performance platform where people can express themselves with music, singing, dancing, videogame playing, and more.

He is also working with MidiCi, a fast-growing chain of Neapolitan pizza restaurants; Agape, a LinkedIn meets Facebook start-up for the faith community; and Jesus Unit, a film production company focusing on faith-based, multimedia properties.

John credits God and hard work for his success both on and off the court. "God put that desire in me to adore basketball. I had to have a strong love to work out three times a day, lift weights, run the track, do thousands of drills, and take thousands of shots. It takes commitment and sacrifice at a very young age to make it to the NBA."

Growing up without a father, he had strong influences in his life, in addition to his mother. One man was in the military and instilled in John the importance of discipline. One woman focused on education and exposed him to different things in life. And then there was the person who put a basketball in his hands when he was eight years old.

"Nothing matched my passion for basketball. I based everything around basketball from my happiness to my self-worth. If basketball was going well, everything was going well. And God knew that and used that."

John loves basketball because it is an individual as well as a team sport. "You can go to the park, work on your game, and imagine making that last-second shot. You can't win without a team, but it's very much an individual sport. I had to get up on my own and work on my game."

Transitioning from athlete to entrepreneur has been deeply fulfilling and challenging. His time in the NBA ended unexpectedly. "It was just over. I was alone at my house with no family when I received the news that what I had worked for my entire life, the life I had lived for the past thirteen years, was done. I had mastered the pivot move during my career as a player. Now I was forced to master it for the rest of my life."

Prior to basketball, John's mother put him on a track-and-field team to keep him out of trouble. He hated it because he was not good. "I knew every Saturday I was going to be embarrassed because I wasn't going to win. But I still had to go to practice every day, work hard, and show discipline and character. It shaped me into who I am now—being able to fight through and continue to work hard."

Being a fighter and disrupting expectations are core parts of building his business empire. "I'm constantly studying and becoming a better business person. I want the knowledge. That was the most exciting part of it—just gaining knowledge and getting better."

Why Firestarters Are Different

I want to be the best. I want to be number one. It's hard for me to accept I'm not number one.

John believes that taking a risk is what makes a Firestarter different. "People want to be safe. They don't want to take the risk to step out and go for their passion. Yes, you can still live a great life by taking the safe path. But, for Firestarters, it's how much risk you actually want to take in life."

He believes taking a risk is linked to the desire to be number one. "Some people say the best position in a company is the number two guy. You get a lot of the benefits from being that high up, but you don't have to worry about actually being the guy that takes all the bad stuff. I think the difference is some people can't be number two. They can only be number one."

He personally is a number one guy. "I still remember who was ahead of me, and it still bothers me. The biggest reason I hated track was because I wasn't number one. I was like six years old, so that's just how I'm wired."

Chapter 29

INITIATORS START THINGS

A good plan implemented today is better than a perfect plan implemented tomorrow.

—General George Patton

Our world sometimes forgets the Firestarters who are Initiators. These are the people who begin things. More people like to think of themselves as Innovators or Instigators. It's just plain sexier. But without Initiators, the world would stand still. They get their feet dirty. They plan. They build teams. And often, they leave their egos at the door.

Entrepreneur and author Seth Godin describes Initiators to a tee: "It's okay. Let your ego push you to be the initiator. But tell your ego that the best way to get something shipped is to let other people take the credit. The real win for you (and your ego) is seeing something get shipped, not in getting the credit when it does."[1]

The Firestarters in this chapter selected Initiators as their dominant type. Here are the stories of the mighty who work constantly behind the scenes, allowing the stage play of life to unfold:

- Ellen Kullman: Former DuPont CEO Trailblazes Fortune 500
- Keith Nolan: Deaf Teacher Fights for Inclusion in Military
- Kim Nelson: *Shark Tank* Winner Shares Love of Homemade Cakes
- David Weild: Legendary Wall Street Executive: Unleashing the Full Potential of Entrepreneurship
- Noel Shu: The "Prince of Luxury" Creates an Empire
- Aleen Zakka: Brand Evangelist Develops Social Media Platforms for Leaders and Politicians
- Joe Edwardsen: *Diners, Drive-Ins and Dives* Pizza Maker Builds Community
- Caroline Tsay: Hewlett Packard Enterprise Executive Shifts into Entrepreneur Mode

- Dr. Anton Berzins: Cross-Cultural Mission Drives School Psychologist
- Heidi Trost: Business Owner Rides to Success by Improving Digital Experience
- Frank Abruzzese and Rosie O'Gorman: Artists Turn Irish Farm into Highly Acclaimed Residence Program and School
- Hezekiah Griggs III: A Tribute

ELLEN KULLMAN

Former DuPont CEO Trailblazes Fortune 500

> *People have to break a big company down into small enough pieces where they can get their arms around it and really drive profits, innovation, and productivity.*
>
> —Ellen Kullman[2]

Ellen Kullman loved her career at DuPont. One of the few women ever to rise to the CEO position of a Fortune 500 company, she has routinely made *Fortune*'s Most Powerful Women list among countless other honors. She considers herself 50 percent Initiator, 30 percent Instigator, and 20 percent Innovator.

She is proof that Firestarters can thrive in large corporations. "I think you have to look at large corporations as if the world is your oyster," she says. "It's about what you are going to make of it. It's not being a small cog in a big machine. It's about how you create."

She finds starting things that bring value particularly invigorating. One of the pivotal points that led her to becoming the CEO was starting a safety consulting business. The company had a reputation as an innovative leader in workplace safety, and Ellen believed this experience and knowledge could grow into a core offering. Eventually, she brought multiple products and services together into a multibillion-dollar safety and protection platform.

Today she is assisting the National Academy of Engineering on educating K–12 teachers to integrate STEM into their classrooms. She is also cochair of the Paradigm for Parity, a collaborative effort of powerful founders and CEOs that have put forth a vision of member companies achieving gender parity by 2030 at all levels of their organizations.

For Ellen, it's about filling the pipeline with female talent at a far greater rate. "We haven't made much progress. There are a lot of companies that want to do the right thing, but they don't know how. If we are not willing to do it as senior leaders who have access to a lot of other senior leaders, then who will?"

Growing up in a large, extended Irish Catholic family, she had a natural curiosity about how things worked. A high school history teacher noticed her strength in science and math and suggested engineering as a career. Her parents had a fierce work ethic and the goal of educating their four children. "My parents had strong personalities. They got it done even when it wasn't easy."

This strength was essential in leading a diverse global organization and in fighting a proxy war at DuPont with activist investor Nelson Peltz. While Ellen won the fight, eventually she stepped down after seven years as CEO because she believed she was a target that would get in the way of the future of the company.

"I considered what the activist was asking to be financial engineering for the sake of a short-term pop in the stock. I did not believe that strategy created long-term value for the shareholder. It was about the greater picture—the company always comes first."

Staying true to principles is a central theme in her life. "I was taught very early on that if you make a decision based on principles—whether people like the decision or not—at least you are consistent in how it plays out. At DuPont, we make hard decisions, but as long as we stuck to our principles we were better as an institution and as a leadership group."

Becoming a business leader was not on Ellen's to-do list when she graduated from Tufts with a degree in engineering. "I wanted to have a good job and a good life. But when I got into business, I became really interested in how to create value and why customers buy from you instead of the competition."

Going to school at night, she earned her master's at the Kellogg School of Management at Northwestern University. Her next stop was General Electric, where she quickly moved up through the organization under CEO Jack Welch. "He was a machine and very impressive. I learned so much about organizational dynamics and how tough decisions are made in a large, complex company."

In 1988, she began working at DuPont as a marketing manager for the company's medical imaging business. She quickly earned the reputation as a turnaround artist for struggling businesses. "At the end of the day, you have to understand how to make money and generate cash because that's what drives your ability to invest and to grow."

She also developed what she calls her closed loop system. "You make a set of hypotheses and measure results. If you're not getting results, you change your actions. Many people just move on, but I believe you need to figure out what is wrong and get to the root cause."

What Extinguishes Her Fire

> *I never worried about my personal brand or what would happen to me. I always felt that no matter what, I would land on my feet.*

As a leader, Ellen's biggest frustration is people with negative attitudes. "Organizations get bogged down by, 'I won't. I can't. We've never.' The phrase I always use is, 'I'm going to focus on what I can control.' If people aren't going to help, I am going to try and push them out of the way."

She relays one story about a disagreement regarding how the public sees science with then CEO Chad Holliday. "I went after this issue five times. He jokingly asked if I had flunked an intelligence test because he had already said no four times. I told him I wanted to go after it a little differently this time. I finally wore him down."

Large companies, particularly ones like DuPont with more than two hundred years of success, have a natural inertia to change, according to Ellen. "It was hard for the organization to see that the future had to look very different than the past. So I had to make a lot of changes in the organization itself."

Why Firestarters Are Different

> *There are leaders who are so invested in the history of how they got there that they protect it. Then there are leaders who take a look at history and say that's great—how are we going to build on it?*

One of the main differences that Ellen sees between Firestarters and other people is whether or not they are growers or stayers. She believes growers understand the need to create a future that is a better place than the past.

"You have to unleash teams that leave the past behind and stop focusing on the way it's always been done. You have to get them excited to try new things. That means engaging them and making it safe by minimizing their personal and professional risk."

KEITH NOLAN

Deaf Teacher Fights for Inclusion in Military

> *It is just a matter of time before the military can finally use deaf and hard of hearing Americans to serve our country and achieve national missions and security. We know it will happen someday, but the question is when.*
>
> —Keith Nolan[3]

If Keith Nolan had been allowed to serve in the US military after graduating from high school, none of what he has accomplished would have happened. Born deaf into a deaf family, this man, who considers himself 50 percent Initiator, 35 percent Instigator, and 15 percent Innovator, was denied the opportunity to serve his country.

The military has always been his calling. Keith's first visit to a recruiter's office was when he was eight years old. At age eighteen, he proudly walked into the local US Navy office to enlist. The recruiter scribbled three words on a scrap of paper: "Bad ear. Disqual."

Neither Keith nor his family let that refusal stop them. His father, the first deaf-born person to be elected to a public office in the United States, suggested he take the issue to Congress and be prepared to go all the way to the Supreme Court. At that point, Keith decided to go to college with hopes of joining the US Naval ROTC. But the answer was always the same: "No, you can't because you are deaf."

After he endured a brief period of giving up his dream after college, his drive, persistence, and passion kicked back in. He went to Israel to interview soldiers because he heard that the Israel Defense Forces accepts deaf Israelis into their ranks who serve essential roles in supply logistics, intelligence, and computer technology. Since 80 percent of the occupational specialties in the US military are noncombat jobs, his mind "spun 180 degrees around."

His alma mater had a new US Army ROTC program, and he was allowed to join. "The door had finally opened, albeit just a tiny crack," he says. "I put my foot right in." Yet the roadblocks continued. Despite being recognized in the top 15 percent of his ROTC class after an incredible first year, he could not get a waiver to continue and finish his training.

This last act of discrimination galvanized him. He took his ROTC experience and did a TED talk that brought the issue to thousands of influencers. He went to more than one hundred meetings in Congress. As the result of his efforts, bills were introduced in both the Senate and House proposing a demonstration program where the military could see the capabilities of deaf and hard of hearing candidates.

The bills never made it out of the Senate and House Armed Forces Committees.

Keith persisted and accelerated his efforts. "We rallied at the White House and the Capitol where some three hundred people showed up. I also picked up incredible mentors who gave important advice and led me to more connections."

Unfortunately, to date he has not achieved his mission. The Obama administration and Congress took no action, relying on a negative feasibility study that Keith feels lacked and misinterpreted important information. Now Keith is focused on bringing the issue to the Trump administration and introducing new bills to the House and Senate.

Still, he feels that a huge change has been made with both the deaf and hard of hearing communities and the general population seeing the feasibility of military service. One example is a cadet program he established at the Maryland School for the Deaf. The cadets learn leadership using military-style structure and discipline as well as advocacy skills to end discrimination in the armed services.

What Extinguishes His Fire

Ever since I first got into the [US] Army ROTC in 2010, it has been an emotional roller coaster. My mom has been the bedrock for me.

Keith says that lack of money, connections, and opportunities are his biggest Extinguishers. Yet they do not stop him. One core reason is his family. "All of my family helped me maintain my tact and professionalism throughout this journey. My mom doesn't get into the details of the politics, but she is always there when the going gets so difficult and frustrating that I get angry. She helps me remember the big picture."

Why Firestarters Are Different

He believes that the differences in Firestarters are two-fold: first, a strong belief in what they are doing, and secondly, whether they have mentors and connections to help them to ignite their lives and make an impact on the world. "It is imperative for people to have a core group who can help maintain the fight every time the going [becomes] difficult."

KIM NELSON

Shark Tank Winner Shares Love of Homemade Cakes

> *I've always felt very comfortable and right at home in the kitchen, and I always knew that there was something very special about the food that my grandmothers and my great-aunt made.*
>
> —Kim Nelson[4]

Some people need to think about the pivotal moment that changed their life. Not Kim Nelson, cofounder and president of Daisy Cakes. Two words. *Shark Tank.*

Her win with mom, Geraldine Adams, on the highly popular show came when Barbara Corcoran took a 25 percent equity position in their cake business in what the real estate tycoon describes as her "best $50,000 investment." The other judges, who all called her product "the best cake they had ever eaten," passed on the deal while cleaning their plates.

With funding, fame, and expert advice, Daisy Cakes went from $88,000 in revenue in 2010 to $602,000 just one year later.[5] Kim who describes herself as 50 percent Initiator, 40 percent Innovator, and 10 percent Instigator became a popular TV celebrity, and her homemade cakes have been featured on QVC, *Wheel of Fortune*, and *The Price is Right.*

Since then the company has grown to a five-million-dollar online cake business. But, despite all this fame and success, talking with Kim is like visiting with a really good friend—she is down to earth, funny, and loaded with Southern charm. Especially when you get her recalling her childhood, standing on a little yellow stool in the kitchen and cooking with her grandmothers Miss Nellie and Miss Nervielee, and her great-aunt Daisy.

When she was ten, she and her mom sold their first cake—a four-layer yellow cake with chocolate icing that she still sells today. After college, Kim owned restaurants, a catering business, and a cooking school. Eventually, the mother of three wanted to leave the intensity of the food service business, so she and her mother went back to their roots and formed Daisy Cakes, using hand-sifted flour, farm fresh eggs, and sweet cream frosting just like her grandmothers and great-aunt had taught her.

But she knew the market in Spartanburg, South Carolina, was not large enough to support the dream she had. She went to Junior League and holiday fairs throughout the country and soon had a national following for the cakes she shipped. When a friend told Kim about a show she had never heard about—*Shark Tank*—she submitted information. More than twenty-two thousand people had applied online, eighty were selected to pitch, and thirty-two actually made the show.

"You only go around once," Kim explains, "and I had absolutely nothing to lose. You know, ignorance is bliss. I didn't make many A's in college, but I did get one in public speaking. I didn't have much to remember in numbers because I had only been a year in business. I didn't really know too much about the show or the judges. So I just did it. The producers just kept telling me to be myself. And that's all I knew how to do anyway."

She says it has been a remarkable opportunity and journey. "I've met so many people over the past six years and stayed in touch with them. People tend to like me. I try to be nice to everyone and am sincerely interested in what they are doing. They then invite me to be on a show or talk to others. It has given Daisy Cakes constant national exposure."

Kim believes talent is one of the key elements that make her a Firestarter, and sharing her cake-making talent with people gives her great joy. "Some people have never tasted a homemade cake that is hand-sifted, made from scratch with no preservatives or artificial ingredients. When people taste our cakes, their eyes roll back in their head."

What Extinguishes Her Fire

> *It's exhausting, but energizing. It's frustrating, but fulfilling. It's stressful, but hopefully more times than not, it's successful.*

At least once a day, running a small business takes Kim up and down on what she describes as the "roller coaster of being a business owner." Cash flow for equipment and repairs as well as retaining good employees are two issues she faces routinely.

"There are days you want to throw in the towel, but then something good happens. At the end, it's about what you *do* to make customers happy and make more sales. And we take advantage of those moments of luck because people really like us—me, Mamma, and everyone at Daisy Cakes."

Why Firestarters Are Different

> *I'm competitive with myself, not just other people. I am always*
> *trying to better myself and make my business better.*

Kim believes that being fiercely competitive is one of the ways Firestarters differ from other people. "We want to be the very best at what we are doing. I want to sell more cakes than I did last year. It's what sets Firestarters apart."

Sometimes, people think that Firestarters are just lucky. "Yes, I am lucky, and some things do come our way like *Shark Tank*. But you must have a great work ethic. You have to work very hard."

Now she is bringing this message of achieving the American Dream to entrepreneurs globally. She got the opportunity from a book interview when the author asked her to speak at his event in New Orleans. "I said, 'Heck yes, and I'll bring cake.' And I did. It turned out to be for one hundred owners of speakers' bureaus."

Luck? Absolutely. Hard work? Positively. Passion? Undeniably. Really, really good cake? Just ask Barbara Corcoran who is still laughing at her fellow *Shark Tank* judges.

DAVID WEILD

Legendary Wall Street Executive: Unleashing the Full Potential of Entrepreneurship

> *We were a nation of explorers going boldly across new terrain*
> *who took great risk . . . without the benefit of modern medicine,*
> *we took wagons across the country . . . and here we are today,*
> *we're afraid that people are going to take risk and lose some*
> *money, to the point in which we've undermined the distribution*
> *that we used to have in small capital markets. . . . That is why*
> *the JOBS Act and other legislation that encourages people to take*
> *risk matters to the future of our society.*
>
> —David Weild[6]

David Weild spent the first fourteen years of his career in senior management positions, including Prudential Securities, which later became Prudential

Financial. In the early 2000s, he became the executive vice president for corporate clients at NASDAQ and, soon afterward, its vice chairman. In 2003, he transitioned from that position to CEO of the National Research Exchange, later founding Capital Markets Advisory Partners, which was later renamed Weild & Co. Holdings.

However, he is most notably known for his role in changing the US laws applied to IPOs, having gained experience overseeing more than one thousand equity offerings during his career. His research has been cited by national magazines.

In February 2012, he wrote about rebuilding the IPO market for smaller firms, and a few months later, the US Congress passed legislation to improve IPO financing rules. As a result, he became known as the head of the New Markets Movements that pushed for an end to one-size-fits-all stock markets. The *Wall Street Journal* heralded him "Small-cap IPO evangelist."[7] As coauthor of the 2012 book *Broken Markets* by Financial Times Press, David made the case for the direct link between lower job creation and the lower number of American IPOs.

In June 2012, David testified before the US House of Representatives Financial Services Committee Capital Markets and Government Sponsored Enterprises subcommittee explaining the issue regarding inadequate tick sizes and how they damaged the ability for Wall Street to properly execute and support IPOs.

As a result of his tireless efforts, in April 2012 President Obama signed into law the Jumpstart Our Business Startups (JOBS) Act. It was recognized that the final bill relied heavily on studies conducted by David and his coauthor, Edward Kim. According to *Forbes* magazine, the above efforts led David Weild to be known as the "father" of the JOBS Act.

Talking with David for any length of time, one gets the sense of his genuine humility despite his great accomplishments. He acknowledges that one of our country's greatest challenges in the modern world is forgetting the practices of our forefathers.

"We need to dare and be willing to accept risk," he says. "One of the things that concerns me about the way this country has gone with capital market is trying to take the risk out of capital markets. . . . [T]hat's the wrong direction to go in. People are going to have to lose some money to be able to make some money and to solve the great challenges of our time."

40 percent Initiator, 40 percent Instigator, 20 percent Innovator

> *I'm clearly an Initiator because I always want to push the ball forward and don't mind being the first to bring up something others have been afraid to explore.*

Being an Initiator has naturally made David an Instigator as well. In essence, someone who is gifted as an Instigator sees things as broken when others don't. One of David's true talents lies in seeing what is broken within the constructs of the financial institutions of this country. He saw that something was definitely broken. He identified that the problem resulted not from a market cycle problem, as the majority believed, but a market structure problem. David realized if he didn't step up and find a solution to the issue others continued to brush under the rug, our country would continue to fail. It is this willingness to look for a problem and not be afraid to take on big issues that has made David stand out.

What Makes Him a Firestarter

> *I think [Firestarters] really are uncommonly aspirational. In my case I've been willing to talk to the masses, receive feedback, take criticism optimistically, and get better.*

David states that it takes courage, doing your homework, and staying focused on your message because as a Firestarter, you will get opposition from others. Most Instigators take a lot of criticism, and it's important to be open-minded and listen to others' opinions. That is one of the advantages of going around and speaking to many different individuals. You gain the value of others' opinions, which will either further enhance your message or alter it depending upon new facts. Being able to differentiate between the two without being intimidated to change the right message is what makes a Firestarter great.

What Extinguishes His Fire

> *People who are unwilling to look beyond their own nose can create great peril because they respond too late to challenges that were foreseeable.*

Not only did David realize the financial issue that would grip the United States if left unchecked, but he wouldn't stop talking until the right people stepped up and took notice. When he first started talking about the crisis as he saw it, people at both bracket firms described him as a "whack job," stating he didn't understand that the world was in the middle of a market cycle.

However, he didn't give up spreading his message, and eventually, when enough time went by, people started acknowledging that David was the "only one who really understood this problem . . . and more and more people started paying attention." David describes that it took five years and about thirty speeches before anyone willingly gave him the podium to speak.

Whenever and wherever he was invited, he showed up and gave his message. He hit pay dirt as a result of this tactic. Someone at the FCC had seen his work, thought it was interesting, and invited him to speak in place of Marion Shapiro at the American Bar Association Security Regulation Watch.

That engagement led him to a Security Regulation Institute luncheon, an engagement where all the important regulators showed up every year. This time, he was in the same keynote where Marion Shapiro had been the previous year, speaking in front of Mary Jo White and her husband, John White, head of the security practice at Cravath, Swaine & Moore. He was reaching the people who needed to hear his message because he didn't give up.

Why Firestarters Are Different

> *When the trains come off the tracks, that's when people like me sit back and say, 'What's going on here?' . . . I think we have done at best half the work we need to do. We still have some looming risks and problems.*

The difference between those who are Firestarters and those who aren't boils down to those who chase the puck instead of skating to where the puck is going. Staying focused on the problem instead of looking ahead toward the solution keeps some individuals held down and classifies others as true Firestarters.

NOEL SHU

The "Prince of Luxury" Creates an Empire

Everyone is given the same amount of time every day. If you are not going to do something that stands out, what are you using your time to do?

—Noel Shu[8]

Noel Shu is a self-made millionaire with a passion for and expertise in luxury and China. At twenty-seven, the Firestarter defines himself as 70 percent Initiator, 15 percent Innovator, and 15 percent Innovator. He has already constructed a business empire in jewelry and wine and is currently expanding his reach into the fashion, food, real estate, and entertainment industries.

Catering to a clientele that includes billionaires, celebrities, and royal families, the "Prince of Luxury" was managing partner and head sommelier for Prodiguer Brands, which sold the most expensive single bottle of champagne in the world for $1.8 million.[9] He has since started his own luxury wine and spirits company, Un Joyau Majestueux, and was the official wine partner for the 2017 Daytime Emmy Awards. He is also the president and chief luxury officer at NJS, which creates custom-designed jewelry.

Noel's reputation and accomplishments earned him a VIP invitation, alongside Bill Gates and Jack Ma, to the 2015 visit by the president of China to the United States. Author of *China Through a Glass of Wine*, he invests in emerging opportunities in China as well as Chinese business models looking to launch in the United States. His film production company, 1768 Entertainment, creates documentaries and suspense movies.

Noel believes that lots of people have great ideas, but they don't act on them. "Regret does not come from taking action. It comes from not taking action. Who knows how many Apple or Alibaba ideas are out there that just died with the person because they never acted upon it?"

His desire to excel and his business savvy are rooted in his childhood as the son of entrepreneurial immigrant parents who were among the first importers of car engines and auto parts from China. Accepted into every Ivy League school where he applied, Noel made the unusual choice to go to West Point.

"If I went to Yale, Harvard, or Princeton, there wasn't too big of a difference. But West Point was something different. It was about being not only smart confident, but also physically confident. That was what I was chasing—to be a cut above the others."

While he loved being a Special Forces combat driver, he was concerned about the long-term feasibility of a military career because of the regimentation, salary caps, and physically intensive aspect of the work that would eventually have relegated him to a desk job. He left and, at the advice of his father, enrolled in the Gemological Institute of America.

"That was the starting point of who I am today. I began associating with a whole different circle of people and going to auctions to look at diamonds. I remember one of the first auctions I attended where the cheapest bracelet was $100,000. A colonel after twenty years of service was probably getting a salary in the low $100,000 range. That was a huge eye-opener for me. I realized that I needed to do better because the world is so vast out there."

His entry into the world of wine rose out of an awkward moment early in his career. He was invited to a private viewing of a jewelry collection on Long Island. "The owner of the estate asked me what I thought of the wine I was drinking. I embarrassed myself because I lacked the knowledge to at least appear intelligent."

He went home and immediately signed up for a program with the International Culinary Institute (because taking action is what Initiators do). Wine became his passion—the whole sensory experience as he combines his deep knowledge of wine with his jewelry design skills to create extraordinary bottles. "Every single impression is important. The actual experience starts when the wine catches your attention."

Noel surrounds himself with people who are skyrocketing in their careers. "What helps keep me motivated every day is that everyone is constantly getting better. When you are surrounded by the kind of people I know, how can you sit on your couch?"

Wealth is also important to him. "Some people take it to the extreme. They think that money is absolutely everything and forget friends and family in pursuit of profit. That is not me. Money is important because if you do not have cash or capital, you can never pursue your dreams and interests."

What Extinguishes His Fire

Sometimes you need to step away from work, leave everything, and come back with a cooler head. You come back doing half the work with twice the amount of results.

Noel believes that his biggest Extinguisher is when he becomes so consumed by the work that it causes him to plateau and lose sight of his original goal.

When this happens to him, he takes a step back and focuses on something else. "I will literally shut down for an entire day, fly to Miami, take a vacation, and then come back rejuvenated to start all over again."

Why Firestarters Are Different

> *I think Firestarters do something, profit from it, and move on while not giving up the pervious venture. They continuously seek something that is different with more profits and more knowledge.*

Noel believes that true Firestarters build empires—people and companies like Jeff Bezos and Amazon, who started as a distributor of products and now create music, tablets, phones, and films, influence him.

"Many people find one thing they are good at and then perfect and profit from it. Lots of smart people out there have great ideas and a specialized craft that they keep practicing. Firestarters are different because they keep building and diversifying. Those are the people and companies that make an impact."

ALEEN ZAKKA

Brand Evangelist Develops Social Media Platforms for Leaders and Politicians

> *I studied the market closely and realized that there were very few 'social CEOs.' Lots of them were avoiding social media. I believed it was a mistake and that I could educate them on why it is so important.*
>
> —Aleen Zakka[10]

When Aleen Zakka tells her story, there is a moment when she pauses because the emotion is so strong. "When my daughter was [a] year old," she says, "I left her for a year and a half in Lebanon so I could work in Dubai. I only was able to see her for thirty-six hours every two weeks. My husband was very supportive, but it was so very hard. However, I needed to follow my dream, and that was the only way I could do it."

For the 60 percent Initiator and 40 percent Innovator, her dream has two important and interconnected elements. The first is to create a highly successful business that enables leaders to build authentic and powerful personal brands on social media. She is accomplishing that through her work with high-profile clients as the founder of Net2Work Solutions.

The second is to empower Arab businesswomen and mompreneurs. Speaking engagements and a motivational series on *Nafizati*, a premium video-based digital magazine for Arabic-speaking women, are helping her fulfill that goal.

Aleen grew up in war-torn Beirut. "When you grow up in a war, you don't think about the future. You think about whether or not you will see your family tomorrow."

The war ended when she was six, but it seriously affected the educational systems, real estate, infrastructure, and the economy.

Her family owned a flower shop, and Aleen has been working since she was a teenager. Her mother was a strong role model, raising her with very little money but making sure that she got the best of educations. Her fellow countrymen and countrywomen also provided inspiration. "The Lebanese people are fighters. We fight for a better life. Because of this, it is easy to succeed within our country. But the challenge is achieving success on a regional level."

She quickly saw that digital marketing was her forte. An early adopter, she had one of the first Facebook accounts in Lebanon. She employed social media to build a solid following on Twitter, LinkedIn, and YouTube. She also broadened her knowledge of other cultures by attending school in France, working in Dubai, and traveling throughout the world, enabling her to speak English, Arabic, French, and Spanish.

Highly strategic in both how she builds her own business and that of her clients, the thirty-four-year-old businesswoman studies the profiles of each person she connects with. She believes strongly that the way to build an authentic social media presence is to connect person-to-person. "Social media is about mastering how to connect with emotion and interest. It is about asking people how you can help them, not about how you can sell them. It's about building trust."

Knowing how to start something is one of her biggest strengths. "I meet so many women who have great ideas. They just don't know how to get started. I'm able to take what I have learned and provide them with the tricks and tips. If you don't start, you cannot move forward."

Going to Dubai is one prime example of Aleen being an Initiator. "I went to Dubai because I needed the connections. I needed to meet face-to-face with people so they would trust me. It gave me the opportunity to listen to so many influencers and understand what was working and [what] was not."

Aleen is intensely goal-driven. Every three months she evaluates what she has achieved and then moves on to more goals. "I love to fuel my passion with new ideas. I ask myself every day how I can achieve a new challenge. That's what empowers me."

Why Firestarters Are Different

True Firestarters make an identity for themselves.

Not surprisingly, Aleen believes the difference between Firestarters and others is that Firestarters actively make an identity for themselves. "When you are an employee, you are always an employee. You don't see the need to create your own identity. But when you are a business owner or a leader, you need to look at it in a different way. You have to have a clear identity."

She feels that she was born to help people and companies create their brands through social media, and her company is the right platform. "I want Net2Work to be top of mind for all CEOs and leaders in the region. And I want my accomplishments to allow me to influence and empower Arabic women."

JOE EDWARDSEN

Diners, Drive-Ins and Dives Pizza Maker Builds Community

I'm a pretty basic guy. The fun stuff is all about making things happen. It's not much more complicated than that for me.
—Joe Edwardsen[11]

Joe Edwardsen finds happiness in moments. Making a pizza. Playing with his daughter. Hanging out with his wife. Chatting with customers. But behind the happy guy exterior is a fierce Firestarter who has worked hard for his beloved adopted city of Baltimore, been integral in revitalizing a boarded-up neighborhood into a thriving arts and entertainment area, and built a reputation as one of the top pizza makers in the country. It's not been easy, but Joe wouldn't have it any other way.

A frequent winner of national and regional awards by both food critics and pizza lovers, his pizza joint and bar, Joe Squared, was featured in the "Best

Pizza" episode by Guy Fieri on *Diners, Drive-Ins and Dives* and on the Bucket List of Pizzerias to Visit before You Die by *First We Feast*. He makes a thin, coal-fired sourdough pizza with some crazy mixtures of toppings like crab and bacon. And, yes, the pizza is square.

But what the 40 percent Initiator, 30 percent Instigator, and 30 percent Innovator is most proud of is building a gathering place for the community through great food, drinks, music, art, and a staff that thinks of Joe Squared as a second home.

One of the more poignant moments was during the protests in Baltimore during 2015. The restaurant was shut down because of a city-imposed curfew, but Joe Squared employees volunteered to hit the streets handing out free pizzas. A month later, Joe received an email from a soldier in the Maryland Army National Guard. Here is an excerpt:

> I can't quite figure out how to explain how surreal it was to be standing in the city that I love and lived in with full gear, a weapon, and ammunition. People would come up to us and just scream, just scream at us for being there. We didn't know these people, no idea who they were and what they've been through but the amount of rage and frustration on their face was incredible. But [we just] had to stand there and take it.
>
> On day two of us being at City Hall, I remember seeing three younger kids on the other side of the picket fence. They were screaming, really screaming at me and the other soldiers. . . . Then two older gentlemen came over and tried to talk to the kids . . . . They ignored the men and kept yelling.
>
> It was then that I saw a young man and woman come over in front of me with pizza. Not only were they bringing pizza to the soldiers and police officers protecting town hall but they were giving it to the people who were on the OTHER side of the fence too. . . . But then something else happened. The two men guided the kids to take some pizza, which they did happily. So then the two nice gentlemen and the kids were now eating pizza together. That's when the two guys started to talk to the kids. About the situation, what was ACTUALLY going on, how violence is not the answer, etc.
>
> It really got through to the kids. By the time they were done with their pizza, they had put their signs down, apologized to me, and gave me a hug before they left. Honestly, that was the most inspiring and faith restoring thing I have ever witnessed and it was all because of Joe Squared Pizza. The world just needs more pizza.[12]

What Extinguishes His Fire

I can't stand greedy people and greedy companies. It completely baffles me why some people just feel the need to screw over other folks.

Joe believes that acting in a generous manner is not only the right thing to do, but also good for business. So he is perplexed and disheartened when corporations and individuals try to hurt his company in ways that he views as unethical and greed-only motivated.

"Running a small business is really hard," he says, "and the restaurant business is one of the toughest. So when you run into a company or individual who is trying to stomp on you just because they can, it can make you lose heart. But I have employees who depend on Joe Squared and a community that has taken us into their hearts. We just keep it going."

He also lives his values through actions. Joe Squared takes no commission on the sale of artwork that hangs on its walls. Employees share tips to encourage good service and collaboration. The bands that play in the restaurant's downstairs music venue take full proceeds from the door. Monthly Do Good Pizzas are named in honor of local nonprofits with proceeds donated.

In return, the community has supported him through some tough times. For example, when the building he was in started collapsing around him because of landlord neglect, a GoFundMe campaign raised $40,000 so he could move down the street to a new location.

Why Firestarters Are Different

I don't think I am more of a Firestarter than the guy who is washing dishes in my kitchen to support his family back in Africa.

Not surprising, Joe sees Firestarters in many people. His staff is made up of immigrants building new lives in America, young mothers going to school, artists and musicians whose passion for their craft has them waiting tables so they can do what they love, and young men and women trying to take their families out of poverty.

"I have no use for people who don't work hard. I have no use for people who think they are entitled to handouts or don't take care of their own. But most of the people I see are really trying to make better lives for themselves. And so many show such courage and resiliency. I'm going to do what I can to help them."

CAROLINE TSAY

Hewlett Packard Enterprise Executive Shifts into Entrepreneur Mode

I started to explore and think of my career in ways that were more than just that day-to-day job, and it's provided me so much more opportunity.

—Caroline Tsay[13]

Caroline Tsay has a résumé that doesn't stop. Hewlett Packard Enterprise (HPE). Yahoo. IBM. Stanford. Her drive and accomplishments have earned her official titles like vice president and general manager and the unofficial title of Firestarter.

Now she has turned entrepreneur, starting a company called Compute Software, an enterprise software as a service (SaaS) company that dynamically optimizes and automates cloud resource decisions.

"I have often operated in entrepreneur mode in large corporations," Caroline explains. "I truly enjoy putting all my energy into driving growth and building high-performing teams. Now I get to do that with my own company. I couldn't be more excited about my new endeavor."

Her go-go-go attitude is one of the major reasons for her success. At HPE, she created a new business and platform for offering enterprise software solutions. At Yahoo, she launched consumer internet innovations that drove five hundred million daily visits and $3.5 billion in revenue. At IBM Global Services, she developed supply chain, order management, customer service, and channel marketing solutions for clients in the high-tech, travel insurance, retail, and medical device industries.

She has been recognized as the National Diversity Council's Top 50 Most Powerful Women in Technology and *Silicon Valley Business Journal's* 40 Under 40. She also serves on the board of directors for several public and private companies.

She credits her experiences at HPE as providing a strong backdrop for her growth as a Firestarter. "We built something great in such a short period of time. So much of my team and organization had an influence on my thinking and my mode of operating."

Today her mission is to recognize the balance between her personal and professional life in a way that allows her to feel happy and fulfilled. She is enthused and dedicated to her new company and also wants to enjoy life with her new husband and eventually have a family.

As 70 percent Initiator, 15 percent Instigator, and 15 percent Innovator, Caroline wears the Initiator label proudly. Her approach is about taking thoughtful actions to make sustainable differences in an organization's success. Her almost four-year tenure at HPE is a prime example.

Instead of trying to fix a traditional sales and marketing organization, she built something new to focus on end user audiences coming through an online channel. As a result, she transformed the way HPE designed product experiences and sold software to technical end users.

"Being an Initiator means that you figure out the priorities, trade-offs, and decisions to make to satisfy your customers, market, and stakeholders in your enterprise. You see all the parts and put them together to execute on a vision and plan. Nothing thrills me more than the challenge of dealing with that complexity and thinking, working with a team, and watching our collective progress accelerate a business."

What Makes Her a Firestarter

> I'm not competitive with other people. I'm competitive with myself.

When Caroline sets a goal, she makes it happen. "I will take away as many road-blocks as I can to make it happen. I will take the steps to go and get it done."

What Extinguishes Her Fire

> I think so much of how I thought about my life was that it was all dependent upon me to make it happen versus opening up the world to say there are people who can help me.

One of her key lessons from working at HPE and also with an executive coach is to ask for help and to help others. "I recognize that my success is really dependent upon other people too. I've opened up to the world, and part of what makes me happy and healthy is helping others. And I also believe that at some point helping others will help me in some way, and I can use all the help I can get."

Why Firestarters Are Different

At the end of the day, you know where your north star is, and you believe you're doing the right thing by pursuing it.

Caroline believes that Firestarters have conviction. "It's that true long-term conviction that makes you go and pursue that path. All the noise and influences around you don't take you away from that conviction because you believe it's truly the right thing to do."

DR. ANTON ROBERT BERZINS

Cross-Cultural Mission Drives School Psychologist

You don't call over four thousand people every year and spend three months on the phone, hours every night, if you're not committed to your mission.
—Dr. Anton Robert Berzins[14]

Mission is a big word for Dr. Anton Robert Berzins. He is the cofounder and director of the only mental health cultural immersion program available to graduate students in psychology, social work, and education. Ecuador Professional Preparation Program (EcuadorPPP) serves needy Ecuadorians who require psychological counseling by connecting them with professionals who want to improve how they work with people of Hispanic/Latino backgrounds.

Anton's mission was forged in early childhood, growing up with a father from Latvia and a mother from Trinidad and Tobago. His immigrant parents believed strongly that no one was better than anyone else and no one was below anyone else, and they instilled that value in their son.

"I was blessed to have two parents that not only exhibited a resilience mindset," Anton says, "but also best exemplified the term 'true grit.' They believed that we are the creators of our destiny, and they reminded my siblings and myself to never take anything or anyone for granted."

As 75 percent Initiator, 20 percent Innovator, and 5 percent Instigator, he initially wanted to become a professional soccer player. Although his love for the sport never dwindled, when he gained acceptance into Columbia University's graduate school psychology program, he made the difficult choice to pursue an academic career over an athletic career.

In graduate school, he attended a cultural immersion program in Costa Rica. This program was created by and run by Dr. Thomas Oakland, who is a world-famous school psychologist who later became his mentor and the key force in him creating EcuadorPPP. Anton learned the importance of mentoring and the positive energy received from guiding people on their path from this professional connection.

Anton started his ten-year-old organization at the age of twenty-five. As an Initiator, he did not care if he failed. It was all about beginning something and giving it his all. "My youth soccer coach always reminded my teammates and myself that we miss 100 percent of the shots that we don't take."

Developing real relationships and connections with others are at Anton's Firestarter core. He particularly looks for people who have authority, competence, and compassion so that he can learn from the best. Additionally, he advocates always taking on more. He has a reputation among colleagues as consistently going the extra mile.

Anton also believes in luck. There were several cultural immersion programs when he started his own after graduate school. However, within a few years, they all ended because of the amount of work that was involved. Each time one disbanded, he said to himself, "Wow, what a lucky break. This is a great omen! I have to keep pushing. I have to reach out to more colleagues and develop our program even further so that it's so relevant, and so significant, that we can't be ignored."

He did not take a paycheck for the first five years; all the money was reinvested back into the company. He still does not see family for nearly two months a year. He works thirty hours a week, in addition to his full-time position as a school psychologist in the Great Neck New York public school system.

"You have to ask yourself, is it worth it and am I willing to pay that price? Fortunately for me, my loved ones get it and they support me. They see my focus, and they see that I am on a mission. They understand that I am trying to do something different and important."

Why Firestarters Are Different

It's the decision to act. It's the simple decision to say that you are going to do something and then you actually do it.

Anton believes action is what distinguishes a Firestarter from other people. Both Firestarters and non-Firestarters share awareness of what is important to them. But Firestarters recognize that there is a deadline and that you don't live forever.

"Firestarters don't make excuses. They see challenges as opportunities. And they aren't smarter than you, or more qualified than you, or better than you in any way, shape, or form. Rather, they act. That's it. They keep trying until they succeed. They learn from their mistakes, and they find mentors, resources, support systems, and networks for which to bring their idea into fruition. You can say that they have a spirit, or a positive energy, or a belief in themselves, which says that they won't stop until the mission is completed. This X factor is really pivotal to their success."

HEIDI TROST

Business Owner Rides to Success by Improving Digital Experience

> *I always want to be better, and I am always willing to learn. I have an insatiable curiosity. I just want to know the answer to everything.*
>
> —Heidi Trost[15]

Heidi Trost is a quiet Firestarter until you get her talking about improving the digital experience and horseback riding in hunter/jumper competitions. Interestingly, it is her passion for both precision and connection that are commonalities in both her work and avocation.

As 60 percent Initiator, 30 percent Instigator, and 10 percent Innovator, the former designer turned usability evangelist has created a new methodology that examines language, interfaces, and visuals in order to create virtual experiences that emulate real life. Relying heavily on research and testing, the LIVE Method is currently helping people throughout the world reduce their frustration with bad websites and applications.

She views the market for her company's skills as huge due to high consumer and end user expectations. "I'd say that 75 percent of websites and apps do not provide the end user with what he or she wants and needs," she says. "People are constantly making mistakes because the interface is so bad. This is not only frustrating, but also dangerous in areas relating to health or safety. It's also a huge problem because a bad user experience translates into a bad brand experience."

Heidi believes her blend of creativity and pragmatism will turn her company into a powerhouse—as well as her blend of tenacity and hard work inspired by her mother. "My mom is the hardest working person I know. My father passed away when I was eight, and we were very poor growing up. But she helped me pursue art and ride horses, which was probably the worst possible sport for someone on one income. But she made it happen."

Working in New York City as a designer, she came to the conclusion that she was not a good employee, so she started her own firm. "I want to run the show. I think it is a combination of wanting to do things my own way, speak directly to clients, and choose my own projects."

She also wants flexibility and financial freedom to pursue her other passion, horseback riding. Heidi admits that even though she loves her company, she would give it all up to become a professional rider. Unfortunately, the road to going professional is extremely high risk and very hard without substantial backing, so she created a work life that allows her to ride and compete throughout the country. What drives her is her love of the sport.

"The relationship you have with a horse is so difficult to describe. It's almost like doing ballet with your horse. They are thinking what you are thinking and are an extension of your body. When it is right, it is so right and so amazing."

What Extinguishes Her Fire

When I am at the barn, the horse is the number one priority and everything else fades away. That is my ultimate therapy—horseback riding.

Heidi suffers from OCD and anxiety, which she has had all her life. They can cause her to shut down and question her abilities. Because she is fierce about being successful, her strategy is to stay ahead of episodes through horseback riding and positive self-talk.

"As long as I am one percentage point ahead of anxiety, then I am going to continue doing what I am doing. It is a constant battle. I am not going to pretend it is easy. It is not. But owning my own business is what I truly want to do, and managing the anxiety is just part of it."

Ironically, her OCD and anxiety also have helped make her clients very happy because she tends to constantly over-deliver. "Perfectionism can be detrimental because my definition of perfect often far exceeds my clients' expectation of perfect. I obsess over projects, often spending much more time than is realistic or necessary."

ROSIE O'GORMAN AND FRANK ABRUZZESE

Artists Turn Irish Farm into Highly Acclaimed Residence Program and School

> *Artists are incredible people to be around. They are critical, creative, and curious. They are very inspiring and interested in art as a means of connecting with the world.*
> —Rosie O'Gorman and Frank Abruzzese[16]

There are few occupations where it is harder to make a living doing what you love full-time than being an artist. But Rosie O'Gorman and Frank Abruzzese have figured it out, and they are doing it the Firestarter way.

Together the husband-and-wife team has created a progressive artist-run school and residency on 180 acres of farmland in the county Wexford, Ireland. Cow House Studios has earned an international reputation for its artist-in-residence program for professional artists and is at the forefront of cross-cultural learning and creative exchange programs for children, teens, and adults.

As codirectors, Rosie and Frank are the visionaries, managers, and primary teachers at Cow House. Rosie's younger sisters, mother, and father also work in the business, preparing homemade meals, managing the studios, greeting guests, working behind the scenes, and tending the grounds and family farm.

At first, the couple seems like a study in contrasts. She's a soft-spoken farm girl from Ireland. He's a gregarious Italian from Philadelphia. Her farm has housed multiple generations for 250 years. His home has been in the family for twenty years. She's a painter. He's a photographer. She's a planner. He's a reactor. She's 50 percent Instigator, 25 percent Innovator, and 25 percent Initiator. He's 50 percent Initiator, 45 percent Innovator, and 5 percent Instigator.

But the ties that bind are strong: a deep passion for art, education, and family; a loving confidence in each other; a strong work ethic; a true enjoyment of people, although both were shy children; and an unwavering desire for freedom and adventure as artists and individuals.

The two met at the San Francisco Art Institute while earning their master's in fine arts. After falling in love at a study abroad program in Venice, they slowly evolved their shared dream to create a self-sustaining art space. They had notebooks filled with drawings and ideas. But they had virtually no idea how to start a business.

"We had the desire, but we didn't know the way. We were artists with no financial training. That was a steep learning curve," says Frank. "My brother-in-law helped us write a business plan. We spent a year doing market research and legwork. We didn't know how to talk to a bank, and we had to get planning permission. Ignorance was bliss. If we knew every step when we started, that would have been a terrible thing. It was much better to take it in small chunks."

They started with a summer program for teenagers and a residency program for artists. The residency program was particularly important. "We have so many interesting and creative practitioners. We get to see their creative process. That informs the way we teach, and the way we move the studio. You can't be a contemporary art space unless you are tapped into the contemporary art community. If we didn't have that, we would get very stale," comments Frank.

Their programs for students appeal both to young people who are building their portfolio for art school and the 75 percent who love art but are pursuing other careers. "The lessons you learn when you try to make art are applicable in so many ways. Much of what you do in school is learning and regurgitating information. There is not a lot of critical thought. As an artist, being faced with a blank page is a daunting thing. So teaching kids methodologies about how to start is really valuable. Our students and their parents have called it transformative," explains Rosie.

What Extinguishes Their Fire

This place is a great platform to experiment. We are opening it up in different ways.
—Frank Abruzzese

The first five years of Cow House's ten-year existence was making it stable. After Rosie and Frank got past that exciting and pressure-filled period, they found themselves in a routine. Frank, in particular, felt restless, which is part of his natural personality: "It felt like the fuel was burning out."

So the couple reassessed everything they did. They began delegating more. They split responsibilities according to each other's strengths instead of being joined at the hip. They invited an outside curator for the residency program. Rosie began connecting with local schools since working with children is one of her loves. Frank started a second master's program in art criticism and theory.

And most importantly, they started consciously carving out time to do their own art. "We had to make our own work a priority," says Rosie. "You can get lost in day-to-day stuff."

Why Firestarters Are Different

It's being aware of what is in front of you and taking hold. You can't just sit passively and wait for something to happen.
—Rosie O'Gorman

Both Rosie and Frank have a clear, shared vision and similar values in life. They believe this is important as a Firestarter couple as well as being deeply connected to what you are doing and a willingness to take risks. As artists, they both take risks continuously.

Frank compares being a Firestarter to going "off the book" in chess: "Professional chess players can memorize thousands of plays from great matches. That's playing on book. When two really great chess players get together, they play on book for a while. But there is a moment in time when they go off the book and do a move that hasn't been done before. They have to go off book to win."

For both of them, there is no better way to live their lives than at Cow House Studios because as they say, "You don't *decide* to be an artist. You just *have* to be it."

HEZEKIAH GRIGGS III

Venture Capitalist, Inspirational Speaker, and Former Teenpreneur (1988–2016)

Raoul Davis, one of the authors of this book, will never forget the call he received December 22, 2016, that his former client and long-time friend Hezekiah Griggs III had died in a car accident. Hezekiah was twenty-eight years old. This profile is a tribute to a true Firestarter who in his short life epitomizes an Initiator with the ability to adapt to the roles of Instigator and Innovator.

Equally important, Hezekiah is the personification of a strong conviction that so many of the Firestarters we have engaged: start something today that makes a difference because life is amazingly fragile, short, and precious.

Fig. 29.1. Hezekiah Griggs III.

Hezekiah Griggs III's name means "He who God has strengthened." Described as "America's Youngest Media Mogul" and a "wunderkind," Hezekiah started his first business at age seven.[17] He was considered one of the most successful young entrepreneurs in the world and endorsed by leaders like former president Bill Clinton.

Among his many responsibilities, Hezekiah served as chairman of the Hezekiah Griggs III (HG3) Foundation, the Initiative for Innovative Leadership & Entrepreneurship, HG3 University, the Youth in Business Development Program (YBDP), and the YBDP National Forum.

While his résumé was impressive, what set him apart was the inspiration he ignited every time he spoke. In every situation, he transformed his age from a potential Extinguisher into an Accelerant as people heard his brilliant insights delivered with professional poise and a baritone voice like a young James Earl Jones. Just read these words (slightly paraphrased for grammatical purposes, from a panel discussion at the First Baptist Church of Glenarden):

> A lot of us think all Dr. King did was dream. But it takes a lot of fortitude to walk down the street and know that every second you are breathing could be your last. That's not dreaming, that's action. That's provoking leadership. That's servant leadership.
>
> One of my favorite Dr. King quotes is when he said, "If you want to be important, wonderful. If you want to be recognized, wonderful. If you want to be great, wonderful. But recognize that he who is greatest among you shall be your servant. That's a new definition of greatness."
>
> In order to be great, we all must serve. Our subjects and verbs don't have to agree to serve. We don't have to know Plato and Aristotle to serve. We don't have to know the second theory of thermal dynamics and physics to serve. We only need a heart full of grace generated by love and all of us can be that servant.
>
> And so if we don't teach the people and if we don't lead by example, then what you have are communities where young people think Dr. King was just a dreamer, and other people think that people like me don't have the potential to be who I am today.[18]

Hezekiah was fueled by the idea of making at least one life better each day. He believed young people need hope, and the message he often shared with them was powerful, yet simple. "Be fast. Be smart. Be young and be strong. But don't forget the future in order to take advantage of today. Always be thinking of tomorrow and the next day and beyond."[19]

He lived by and shared his eight principles of life in order to ignite others, especially young people.

1. **The Principle of Perseverance.** Life is a beautiful struggle.
2. **The Principle of Passion and Purpose.** Truth is hidden in wisdom.
3. **The Principle of Strategy.** Where there is no vision, there is always failure.
4. **The Principle of Counsel.** Sometimes in life, you don't know enough to know.

5. **The Principle of Professionalism.** Self-respect permeates everything in life.
6. **The Principle of Support.** Leadership is obtained through the service you give to others.
7. **The Principle of Competition and Service.** Life is not a game, but there is competition.
8. **The Principle of Investing in Yourself.** The greatest investment is when you invest in yourself.[20]

The words of his grandparents in the program for his funeral offer the best insight into how he was able to accomplish a full life in so few years and leave lasting Emissions that will ripple through all those he mentored, heard from him, or know his story. "In spite of the petty, mundane, and trivial realities through which everyone must thread, he saw beyond these things and beheld the brilliant lights of accomplishment and fulfillment that lay ahead in his life."[21]

If Hezekiah had been interviewed for this book, we believe he would have told us that what makes Firestarters different is that they understand there is pain, recognize it, expect it, and decide to go forward anyway. Hezekiah's words are inspiring:

Through all my pain, I'm still standing. Whatever someone might be going through, there is a way to turn your life into the positive energy you need to reach the mountaintops of your desires. My life is living proof of what a determined individual can do [to] shape the life he or she wants to lead.

I truly believe that you have the power to chart your own path in life, no matter your situation. So by simply committing to making your world better, through the choices you make and the decisions you come up with, you are already doing your part to set the world right. A better me equals a better world.[22]

BURNING CONCLUSIONS

The Firestarter Framework

Innovators	Instigators	Initiators

Firestarter Types

Igniters	Fuels	Accelerants
Situational Motivators	Environmental Resources	Context-Driven Action
Freedom in the Moment	Prestige	Mission Focus
Tapped Talent	Opportunity	Collaboration
Mastery Mindset	Wealth	Constructive Competition
Driving Passion	Luck	Sweat Equity
Global Introspection	Social Connections	Support Seeking

Emissions

Supporting Factors

Cognitive Convergence	Avoidance of Extinguishers

Chapter 30

SUPPORTING AND EXTINGUISHING FIRESTARTERS

T hroughout this book, we have focused on how you can maximize or discover your own Firestarter potential. But, there is another important question: what is your impact on the other Firestarters in your life? You have a choice: support the person and help spread his or her fire or work to extinguish his or her flame.

Supporters keep the flame alive. They look for ways to supplement the Firestarter's own Igniters, Fuels, and Accelerants and to prevent Extinguishers from snuffing the flame. Extinguishers do the opposite. They are the human manifestation of the challenges Firestarters face. Extinguishers in their human form add obstacles to the Firestarter's path.

Firestarters must successfully align themselves with Supporters. They must also avoid the crushing blow of Extinguishers. It's not a tough choice to make. But there are nuances, and sometimes, it is not directly obvious who is a Supporter and who is an Extinguisher in your life.

Why? Extinguishers, while they damage your potential, can have honorable motives. They may think (and may seem like) they're helping you.

FOUR EXPERTS GIVE THEIR VIEWS

Before we talk in depth about the types of Supporters and Extinguishers, we'd like to take a side trip and share with you the advice of four people who in our view have breadth and depth of experience that gives them a deep understanding of how to support Firestarters.

Joe Mancuso is the founder of the CEO Clubs, a forty-year-old organization of CEOs with chapters throughout the world. Mancuso believes that "most Firestarters are a little crazy."[1] They often buck trends in ways you wouldn't expect, and understanding their unique quirks is part of supporting Firestarters. Here are a few examples Mancuso shared with us:

- Michael Bloomberg, former New York City mayor, told Mancuso he takes the same approach to employees who resign as he does with ex-wives. He never wants to see them or talk to them again.
- Herb Kelleher, former CEO of Southwest, told Mancuso he smokes two packs of cigarettes a day and doesn't worry about dying.
- Mancuso recently launched CEO Clubs of Pakistan with membership packages ranging from $2,500 to $10,000. He asked people why they bought the $10,000 package, and they said, "[I]t was the most expensive one."

These examples make Mancuso's point poignantly. "Many people end up unintentionally becoming Extinguishers because they are unable to follow the logic of many Firestarters. They push back, which frustrates the Firestarter, stunts their ability to grow, and leads to frustration or high turnover. While most Firestarters love being challenged, they don't like the challenge to stem from the other person's lack of imagination."

John Sculley, the former CEO of Apple who is profiled in this book, shared his thoughts on what some people call Steve Jobs's perception of the world—a reality distortion field. He said Jobs believed that reality should bend toward his ideas on numerous occasions and that it was the source of his greatest strength and challenge.[2] It made his ideas at times awe-inspiring and simultaneously made him the biggest pain many people ever worked with. One can look at the portrayals of Jobs in several theatrical releases and see he was nearly always indifferent to how he was being perceived.

Counterintuitive thinking and apparent callousness toward receiving feedback can make working with highest-performing Firestarters a daunting task. What Sculley and Mancuso have discovered is that it often isn't for the faint of heart. To be a Supporter of a Firestarter you have to be willing to buy into his or her bigger vision and allow yourself to stay bought in. You can avoid becoming an Extinguisher by not getting obsessed every time their tone may sound condescending or the direction they want to take stretches far outside the box.

As the founder of the Women Presidents' Organization (WPO), Marsha Firestone has seen literally thousands of women-owned businesses take their highly successful businesses to the next level. Eighty-two percent believe WPO helps them manage their business concerns through support, empowerment, inspiration, and the ability to tap into the wisdom of their peers.[3] Firestone believes Firestarters need support from people who have experienced similar opportunities and challenges. That's why confidential and collaborative peer-learning groups are at the core of WPO.

Ellen Kullman, former CEO of DuPont, believes supporting Firestarters is helping them have a clear vision along with measureable actions. One place this is evident is her current work with Paradigm for Parity, where member companies are working to achieve gender parity by 2030 at all levels of their organizations.[4]

This group is supporting each other to achieve a common vision by committing to five measurable actions: (1) initiating unconscious bias training; (2) setting targets for share of women in top jobs; (3) measuring the annual progress the company makes toward those goals and sharing them with others; (4) evaluating employee performance on results, not face time; and (5) providing sponsors for promising female employees.[5]

SUPPORTERS

Don Quixote has Sancho Panza. Johnny Carson had Ed McMahon. Batman has Robin. Kim Kardashian has Kris Jenner. From superhero sidekicks to momagers, Supporters fan the flame of impact. They stand by the side of the Firestarter in tough times and find ways to keep the passion alive.

Why do they do this? The motives differ from person to person. Some have a genuine interest in spreading the other's flame. Others see a path to self-aggrandizement. Regardless of motive, they see a Firestarter, and they want to be a part of it.

Through our research and interviews, we identified four primary types of Supporters:

1. Nurturers: people who listen to Firestarters and allow ideas to flourish.
2. Motivators: people who focus on others' passions and work to grow their influence.
3. Illuminators: people who shed light on hard-to-grasp concepts to improve a Firestarter's ideas.
4. Protectors: people who accept a Firestarter's eccentricities and defend his or her path against those who don't understand.

Nurturers

We all know people whom we would describe as good listeners. They seem genuinely interested in what we're saying. They lean forward. They ask questions that make sense. They help to clarify points that we haven't fully fleshed out.

Firestarters need listeners. Not because they need an audience (though some do). Rather, ideas unspoken rarely materialize. Firestarters don't change the world in a vacuum. They need people who are willing to vet their ideas, poke holes, and help to finalize them with additional strength.

Nurturers allow ideas to flourish. They remove ego (as much as possible) and show genuine concern for the Firestarter's inspired direction. They truly want the Firestarters in their lives to succeed, and they show this through active listening and open admiration.

Motivators

In 1519, Captain Hernán Cortés landed in Mexico with eleven ships and just over five hundred soldiers. Their mission: conquer the Aztecs and enrich the Spanish empire (and Cortés). Facing thousands of Aztecs in a military coup would be no easy task. To build commitment, he established a city with his soldiers as citizens (and himself as the elected leader). Cortés feared that many of his soldiers would become frightened or flee in the face of the challenges of conquering a large civilization. As insurance against this possibility, Cortés ordered his army to burn the ships that had brought them to Mexico.[6] His crew likely thought he was crazy, but this was an important step.

Why? First, it gave them something more to inspire the upcoming fight. Those not fully emboldened by the goal of enriching Cortés now had a mission to save their own lives. It also forced interdependence and teamwork in a way that none had anticipated. Every soldier was invaluable. They had to find ways to secure their continued freedom by working together and capitalizing on unique strengths. Within a couple of years, Cortés and his army (bolstered by local allies) overran the capital of the Aztec empire.

Cortés's actions were effective, but not all motivation is so forceful. Think of seminars led by motivational speakers like Tony Robbins. Such individuals find ways to embolden your feelings of competence, confidence, and commitment to yourself (without putting your life on the line). Individuals listening to these motivational speakers find their thought process changing. Tony Robbins has developed a great experiential metaphor for this transition. He guides seminar audiences in "firewalking" in bare feet over a bed of hot coals.[7] This symbolizes to attendees their own ability to overcome fears.

This is what Motivators do. They provide the push you need to discover your own Igniters. Many times, people possess dormant capabilities and confidence. Motivators help them to unearth this potential, learn how to listen to the calling of their passions, and work toward their goals. They "burn the ships" of doubt and blame to help propel others forward.

Illuminators

Sal Khan is a Harvard Business School graduate in the financial world. He has a knack for mathematics and engineering, fortified by an MIT undergraduate pedigree. He is also Bill Gates's favorite educator. After creating YouTube videos to help tutor a cousin, Khan's content started receiving thousands of hits. People recognized Khan's simple explanations of complex mathematical concepts as innovative and fresh. Soon, the Khan Academy was born, providing a large cache of online instructional videos that teachers across the world use to supplement their classroom education.[8]

Khan, a Firestarter in his own right, has a personal mission to be an Illuminator for others. Many great teachers actually serve in an Illuminator function. Think of Dr. Kirk Borne whose story we shared. His personal love of science resonated with and inspired his students.[9] Think of the Greek philosophical triad: Socrates, Plato, and Aristotle who inspired learning and exploration in each other from one generation to the next (and whose philosophies continue to enlighten others).

An individual's idea is a wonderful thing. An individual's idea strengthened by the force of others' intelligence has even greater potential. Illuminators make things better. They help Firestarters learn, discover, and recognize how their innate talents can influence the world.

Protectors

In the Ironman and Avengers movies, Ironman's alter ego Tony Stark is a wealthy capitalist who has a love-hate relationship with his own humanity. He is a genius businessman and inventor, but he often lacks the humility to accept his own limitations. His connection to the real world is embodied through his secretary Pepper Potts. She coyly says things like, "I do anything and everything Mr. Stark requires."[10] And in some ways this is true. In Tony's world, the requirements often involve defense of his own eccentricities, both to himself and others.

Pepper is a Protector. She predicts Tony's actions before he knows what he's going to do himself. She has thick skin and is not offended by Tony's wry sense of humor when others are. She defends and protects his ego from threats, even at the expense of her own best interests.

Protectors appear selfless in ways that others cannot mimic. Their own identity becomes intertwined with someone else's success, and they come to view the other person's accomplishments as their own. They are perfectly content to play second fiddle, so long as the star's solo goes without a hitch.

EXTINGUISHERS

In section 5, we discussed Extinguishers in the general sense. Extinguishers are those things that threaten your fire. At home and work, Extinguishers may be set in motion directly by other people in your life. When this occurs, we refer to those people as Extinguishers themselves.

When your parents question your life choices, they can be Extinguishers. When your boss adds restrictions on how you do your work, he can be an Extinguisher. When your spouse notices a mistake you made but lets you fail, he can be an Extinguisher.

There are four primary types of Extinguishers. The language we use to describe three of them—Limiters, Discouragers, and Punishers—also reflects the language we used when talking about Extinguisher factors in general. These titles are the personification of those elements:

1. Limiters: people who reduce others' Fuel supplies.
2. Discouragers: people who delegitimize the Igniters tapped by others.
3. Punishers: people who shun or ridicule others for their risks and actions.
4. Enablers: people who allow others to mismanage their lives and to succumb to obvious ineptitude.

Limiters

When your Fuels reach limits, your fire can be extinguished. When other people exacerbate the effects of the Fuel limits, they serve as Limiters. For example, when someone refuses to recognize your power, your prestige becomes impotent. Imagine a celebrity's outrage when she asks, "Do you know who I am?" With opportunity, other people can block your chances. Think of the boss who offers a promotion to a favored employee, despite skill set deficits. If you feel that you deserved the promotion, your boss was a Limiter for you.

Wealth is a tricky one. Often, we are placed in the unfortunate circumstances of having to entertain requests for monetary assistance from friends or relatives. For some, this is an easy decision: you don't lend money. Others evaluate the legitimacy of the cry of those in need before acting. When someone refuses to provide financial assistance (whether it is a legitimate or unreasonable decision), that person is a Limiter.

So how can someone limit your luck? By lying or defrauding. Imagine a pollster calls to seek your political opinions. You don't want to be unhelpful, but you also don't want to share your true opinions. So you fudge a little bit. No harm, right? Well, maybe—but you did create statistical noise. Your bad data is a Limiter to the conclusions that prognosticators can draw from the study.

With social connections, your limiting power lies in using the word "no." When friends call asking for a favor, you can deny them. When a new acquaintance seeks information to help land a sale, you can deflect the request. In these scenarios, you refuse to allow others to use their connection with you as a Fuel.

As you can see, Limiters don't always intend to inflict harm. You can have perfectly legitimate reasons for making any of these decisions. But the result is the same. A fire that could have been aided is reduced or eliminated because of choices that you or others make to limit the Firestarter's chances of success.

Discouragers

You may have experienced micromanaging supervisors or controlling parents. What do they have in common? They exercise power to thwart another person's chosen direction. They do this by separating a person from the things that ignite her. Discouragers find ways to block your freedom, talent, passion, and confidence.

They force you not to do freely chosen *constructive* activities. A parent urging a child to stop playing video games is not a Discourager. But a parent who urges you to change college majors from your desired field may be one.

Earlier we discussed how threats, surveillance, negative evaluation, and deadlines stifled passion. Discouragers are the actors who place these albatrosses around your neck. They find ways to make you feel controlled, scared, and anxious when you think about your preferred, constructive courses of actions. Several of the people we have profiled in our book like Dr. Angela Marshall and LaToyia Dennis told poignant stories of Discouragers in their lives and how they overcame them.

We have also explained how role identity motivates action. For example, people identifying as Innovators would be drawn to creative activities. Instigators would be drawn to situations when they can question the status quo.

Discouragers place situational constraints on you that do not allow you to act according to your preferred identity. Think of the highly creative person who is forced to perform rote, tedious tasks at work or school. Or the skeptical person forced to comply with actions that don't make sense to him. The Innovator who is discouraged from being innovative and the Instigator discouraged from questioning will be less likely to reach their potential.

Punishers

The distinction between Discouragers and Punishers is nuanced but important. Discouragers try to dissuade your actions and stifle your motivation in
a mental way. They get in your head. Discouragers *prevent* your undesired
behavior from emerging. Punishers, on the other hand, find ways behaviorally
to drive negative outcomes if you act in ways they don't approve. Punishers
react when you have already behaved in non-preferred ways (though prior
experiences with punishment could eventually discourage you too).

A parent who dissuades you from your desired major is a Discourager. But
a parent who kicks you out of the house or defunds your tuition for taking a
course with which she disagrees is a Punisher. The difference is the intent. Does
she want your action to change, or does she want you to suffer for your choice?

Punishers are difficult people to have in your life. Why? Because they are
usually people you care about or whose opinions matter. You don't care what a
complete stranger thinks about your decisions, do you? Probably not. But you
are preprogrammed to not want to upset those who serve as Punishers. Their
negative reactions hurt, and you may be willing to limit your own potential in
order to avoid triggering their disdain.

Enablers

When people mismanage themselves or act ineptly, you can often see it from
afar. You could choose to thwart the self-harming behavior or choose to enable
it through inaction or even encouragement. Enablers face a conundrum. If
they choose to act boldly to dissuade actions, they may serve as Limiters, Discouragers, or Punishers. But if they let you fail, they can serve the enabling role.

In some situations, being an Enabler is unavoidable. In your mind, you must
either refuse to support a risk that you view as unnecessary or support a risk that
potentially leads to failure. When things have higher levels of risk, the outcome
is less certain or predictable. Thus, if you choose to intervene (or not) you are less
clear in your own mind which direction will yield the most positive result.

In this sense, you often won't know you're an Enabler until the failure
happens. The key is recognizing patterns in your relationship with the other
person. If the person repeatedly overinflates his or her abilities or adopts overconfidence in the face of sure failures, you may ditch the Enabler role and
have a serious heart-to-heart. If you continue as an Enabler after the person
encounters several similar failures, your support becomes untethered to reality.
You serve as a systemic Enabler of bad decisions.

SEEKING SUPPORTERS AND OVERCOMING EXTINGUISHERS

The formula seems simple. Seek and invite into your life those who will support your flame and find ways to limit the influence of Extinguishers. This would be easy if everyone wore a T-shirt describing his or her motives, but reality is not so clean.

A person who is a Supporter in one aspect of your life may be an Extinguisher in another and vice versa. These roles are situation specific, not person specific. Accordingly, you can't always eradicate certain people from your life and call it a day. You must be more discerning.

Extinguishers have power over your decisions, but only if you let them. Supporters contribute to your successes, but only if you invite them to do so. The common denominator is you. You decide who influences your outcomes in any situation.

Your Firestarter potential is intimately intertwined with other people. This is clear from our discussion of Igniters, Fuels, and Accelerants. Firestarters make their impact in large part thanks to the people in their lives who they tap for talent, money, connections, partnership, and advice. Your not-so-simple goal is to find the right combination of people whose contributions to your dreams are most likely to make your desired impact a reality.

SPARK YOUR THINKING

1. Are you able to list the Supporters and Extinguishers in your life?
2. Can you see any patterns in the people you have selected? Do some people serve both roles?
3. Do you have a high number of Extinguishers? If so, did you invite them to serve in that role?
4. Are you an Enabler? Do you see it as a good or bad way to be?
5. Do you believe that it is easy to slip from a Protector to an Enabler? If so, how do you prevent that from happening?
6. How do you find more Supporters for your life?

Chapter 31

UNDERSTANDING EMISSIONS

The great use of life is to spend it for something that will outlast it.

—William James

The science fiction genre in movies and books is often the blueprint for where humankind is going. The genre inspires people to develop or shape ideas they otherwise may not have ever had. For example, the replicator devices in the *Star Trek* television series inspired 3-D printing.[1] Decades ago driverless cars were only a wild dream; today they are a reality. When life imitates art in this way, it is a form of Emissions. Emissions occur in all industries when the impacts of Firestarters develop a life of their own.

So how do Emissions occur? Can they be planned, or do they occur organically? In past generations, Emissions often developed organically; however, technological improvements have allowed greater flexibility for Firestarters to plan for and have a greater role in their own Emissions. Think of Silicon Valley inventors who brainstorm "spin-off innovations" of their products and actively create platforms for such innovation through hack-a-thons and open source coding.

In chapter 28, we presented the profile of career futurist Larry Boyer who spent much of the past few years theorizing around the role of artificial intelligence, robotics, and automation in the workforce. Based on his understanding of data analytics, technology, job trends, and scientific research, Boyer predicts that there will be driverless commercial trucks, driverless forklifts, and system-wide automated service at fast-food restaurants.[2] These are all predictable innovations for which Boyer assists people and organizations. From shifts in the job market to evolving skill needs, people must act to maintain relevance in a world disrupted by innovation. In this sense, Emissions have a price: the need to adapt.

Emissions don't appear overnight. Think about how much havoc and uncertainty the advent of the automobile generated in the early twentieth century. Was the horse-and-buggy industry aware it was headed toward extinction? Did oil and gas companies realize how much their growth and profits would explode? Did auto repair shops immediately begin to rise in prominence? Aside from early adopters and futurists, the answer is no on all fronts. Societal innovation occurred slowly.

While slow, other Emissions are more deliberate. Many don't realize that Rosa Parks's decision to sit in the front of the bus in 1955 was a planned event. The National Association for the Advancement of Colored People (NAACP) had been planning an act of civil disobedience for a few weeks.[3] They believed a woman would be seen as nonthreatening, and they hoped to evoke sympathy.

We mention this not to take away from the power of the Parks story but to demonstrate how Emissions can be anticipated. Nonetheless the NAACP was probably unaware that the impact of their decision would still be written about several decades later. In addition, the small protest of Rosa Parks had an impact that spiraled beyond the eventual Civil Rights Act of 1964 to other legislation that has paved the way for equal protection for women, the LGBT community, and people of multiple ethnic and racial backgrounds.

Overall, we find that Emissions present themselves in four key ways:

1. *Legacy.* A person's story (like Rosa Parks) outlives a single point in time to become a rallying cry for societal change or innovation.
2. *Culture change.* A Firestarter's impact changes the way people interact with and interpret the world. Think of Henry Ford's automobile assembly line or Steve Jobs's push to put control of your entire life in the palm of your hand.
3. *Future innovation.* One action becomes the launching pad for further action and impact. Computer programming has evolved from an individual coder at a single machine to a worldwide network of machine learning. This shift will eventually rock the tectonic plates of reality in ways that we can't even dream of yet.
4. *Inspiring passion in other Firestarters.* Firestarters beget Firestarters. They inspire each other. They mentor each other. They read about each other (you're reading this book, aren't you?). Firestarters seek to learn and understand not just for the sake of knowledge. They start with the contributions of those who precede them and find ways to grow their impact.

In the process of studying and interviewing Firestarters, we've been fascinated by the Emissions they produce. A person's impact extends beyond personality or ego. Firestarters may be motivated by pride; we all are to a certain extent. They may desire wealth or accomplishment of a mission. But the world is a complex, interactive place, and one person's drive feeds the collective drive toward societal evolution. A Firestarter makes an impact, but the extent of the impact determines whether the Firestarter is more than just a blip in history.

SPARK YOUR THINKING

1. What would you like your Emissions to be? Are you acting consistently with this aim?
2. How are you powering societal evolution?
3. Who do you believe are the people in history with the most Emissions?
4. How can we encourage more people to become Firestarters and produce Emissions?

Chapter 32

WHY WE NEED MORE FIRESTARTERS

I cannot stand by while innocent lives are lost!
—Diana Prince, from the 2017 film *Wonder Woman*[1]

In June 2017, *Wonder Woman* made its theatrical entrance to acclaimed reviews and positive noise on social media. The movie, which is set during World War I, offers inspiration and a reminder about both the good and evil of humanity. There are two pivotal moments that struck us while we were writing this book. If you haven't seen the movie, feel free to skip the next two paragraphs to avoid spoilers.

The first pivotal moment is when Wonder Woman and her cohorts enter a British bunker near a war-torn village. Women and children are being killed and enslaved by German soldiers. Wonder Woman bolts into action by herself under heavy fire. Within seconds British troops move in behind her, taking out the enemy and eventually freeing the town. The British soldiers haven't moved in a year, but because of Wonder Woman's actions, they take a risk and change the outcome.

A few scenes later, the Germans launch a new chemical weapon into the town, essentially killing everyone Wonder Woman and the British soldiers have just saved. Wonder Woman doesn't allow this moment to become the Extinguisher that vanquishes her flame. Instead she channels her energy and overwhelms the enemy.

While *Wonder Woman* is a fictional tale, it encompasses the best of the Firestarter reality. It showcases the relentless passion, the channeling of talent, and the ability to avoid or overcome Extinguishers. Lastly, it demonstrates the need for Firestarters to have support.

Each generation has its "Wonder Woman." Mikhail Gorbachev transitioned the Soviet Union despite significant internal pushback. Henry Ford developed the automobile industry, and Harriet Tubman risked her life daily for the freedom of other people. Wipe any of these people from history and the world is a very different place. Think about all the people they've inspired to become Firestarters and how their words, actions, and inspiration continue to shape history.

Being a Firestarter isn't always about an invention that changes the world. It is also about everyday heroes who change lives in their own communities. If you put enough of those people together, collectively they reshape the world. Take a look at the events of Hurricane Harvey in the Houston, Texas, region in August and September 2017. There were countless heroes, people who in a harsh moment stepped up to the plate.

For example, brothers Jonathan and Joshua Evola saw the suffering occurring in flood-ravaged Houston. They were two hundred miles away in Dallas. They could have lain on their couch, just thinking, *What a shame!* However, these two brothers took it a step further. They sprung into action, driving seven hours with their boat to rescue dozens of people. Many others like the Evola brothers came together to help their fellow Texans. At a time when there was a great deal of concern about a lack of empathy in the country. Out of the darkness of all the destruction was the light ignited by the countless everyday heroes who found the Firestarter in them.[2]

As humanity evolves, the problems become more complex and the potential consequences more grave. It took years for World War II to result in the death of a staggering fifty million people.[3] Today, with the advanced delivery systems for nuclear arsenals, that many can be killed in minutes. Innovation has never occurred as fast, yet it carries a deep responsibility.

Having more Firestarters is pivotal. The next generation of scientific advancement will focus on artificial intelligence, automation, and robotics. Truck drivers may soon be without jobs. One day humans may contemplate fusing with AI and downloading their minds into androids. These are things that may seem unfathomable. However, if you explained a smartphone to someone from the 1800s, that person would have thought the same thing.

It will be Innovators who spur these innovations, Initiators who push ideas forward, and the Instigators who fight against the status quo. We don't know what exactly is around the corner. However, we know we'll need Firestarters to be there to innovate, instigate, initiate, inspire, and solve problems. We hope reading this book has helped you ignite your journey or accelerate your path to becoming a Firestarter.

SPARK YOUR THINKING

1. What do you believe would happen in the world if we had more Firestarters?
2. Do you believe that our view is realistic and achievable? Why?

Chapter 33

OUR TOP TEN LIST

After two years of reviewing research, documenting historical examples of Firestarters, talking with hundreds of people, and having weekly debates, we have reached some collective conclusions. With three coauthors, this journey has been intriguing. We are quite different in our ages, backgrounds, expertise, and belief systems. Accordingly, at first we each interpreted the research and interviews through our own lens. We challenged each other. We disagreed. We argued. We even argued about arguing. But our mutual synthesis of this information led to joint conclusions. Unequivocally, collaborating together made our thinking richer, smarter, and more engaging (at least we think so).

So while we and our profiled Firestarters already have shared many insights in this book, we would like to present our top ten conclusions. Our brains actually feel better with lists. They give us a sense of control because they sound definitive. We hope they provide you with a starting point to engage in an enriching and thought-provoking dialogue that changes the way you think and act.

1. We all have Firestarter potential. There is a spark in everyone. It can be nurtured, ignored, or extinguished. Being a Firestarter is not a state of mind. It is a state of action. More precisely, it is a person-by-situation interaction. Firestarters arise in all walks of life, from the stay-at-home parent to the corporate CEO. There is no Firestarter script—no role they have to play. They choose their own lives and determine their own paths.

2. In a very anti-Jeffersonian way, we are not all born equal. We don't have access to the same Fuels. We aren't all driven by the same Igniters. It is easier to be a Firestarter if you have passion, talent, money, connections, prestige, and access to other elements that have been discussed in this book. But mere access to those elements does not make you a

Firestarter. Some people waste gifts that are easily attainable. Others claw and scratch their way to great impact. Firestarters find ways to take advantage of the Igniters and Fuels available to them, despite limits that would stifle others.

3. Most people have a dominant Firestarter type—Innovator, Instigator, or Initiator. Some highly relate to one type; others almost seem like "hybrids" with more balance in all three areas. Most Firestarters also move between their types depending on different points in their life and specific goals and tasks.

4. There are times in Firestarters' lives when they get detoured. They lose purpose, shift priorities, or are burdened by overwhelming personal challenges. They all face Extinguishers, but the strength of their Igniters, the quantity of their Fuels, and the action of their Accelerants push them to high levels of achievement in spite of the obstacles. Firestarters persist. If Thomas Edison had quit after thousands of unsuccessful light bulb prototypes, would his impact be as large? No. If Dr. Martin Luther King said, "I have a daydream, but it's kind of a pie in the sky," would we still be discussing his legacy? Likely not. Firestarters stare down obstacles and work to overcome them.

5. Firestarters have faith in themselves, their mission, and a sense of higher purpose. For many, faith is driven by a religious or spiritual foundation. For others, faith is derived from a mission that consumes their action.

6. Childhood is a formative time for Firestarters. They discover whether Supporters are as close as the next room (e.g., parents) or whether they have to search elsewhere for a situation that will allow them to ignite. You can find Igniters in both positive or negative childhood events and experiences. For example, your parents could nurture freedom, or they could be so controlling that you escape to seek freedom elsewhere. Either way, you find the freedom that supports your journey.

7. Other people are essential to building the largest fires. When others nurture your Igniters, supply you with Fuels, and partner with you to spread the fire, the journey is a lot smoother. Firestarters who try to "go it alone" are more likely to succumb to Extinguishers.

8. Social balance is the key to spreading the largest fires. Successful Firestarters perform actions consistent with altruism, cooperation, individualism, competition, and disturbance of the status quo. Social balance ensures you adopt a rhythm that supports your work.

9. No matter who you are, if you're a Firestarter, someone is going to

think you're a pain in the ass. Why? Well first because you might actually be one! Second, because you have ideas, you share them, and you proactively work toward them. There will be people who disagree. You start moving when others aren't ready to move, and that makes them upset at you for their own discomfort. Firestarters do great things for society—one of them is serving as a scapegoat for ills. Some of your ideas will be bad, some will fail, and some will piss people off. You'll pursue them anyway, and when things fail, people will point their finger at you. Others will enjoy the spoils, produce relatively little, and laugh at your naiveté as you change the world.

10. The world is a better place because of Firestarters. They inspire us, create the future, and find ways to overcome obstacles that others view as insurmountable. It's a Firestarters' world, and we're all on for the ride.

SPARK YOUR THINKING

1. What is your top ten list of insights from reading *Firestarters*?

INDIVIDUAL AND GROUP EXERCISES TO IGNITE YOUR LIFE

The Firestarter Framework

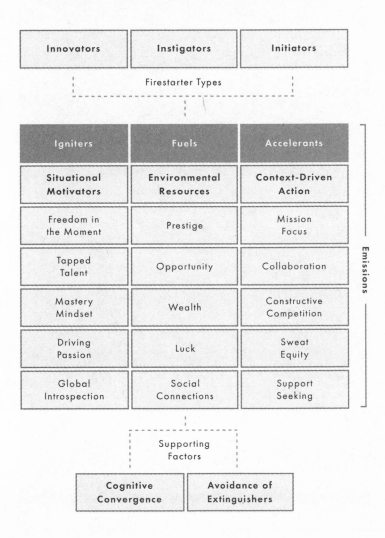

Chapter 34

DEVELOPING YOUR FIRESTARTER STORY

You are on a journey. Some of you have achieved your goals and declared, "Success!" Others are still navigating the roads. Still others need to turn on their mental GPS systems, not knowing where it will lead them. Often, we don't take the time to think about our own journeys in introspective ways. Our lives are usually "go-go-go" and rarely "stop and think."

So we ask you now to embrace that rare moment. Imagine your story is one that has to be told to inspire others and to teach them the wisdom of your experience. We have developed ten questions for you to ask yourself to aid in this introspective exercise. We have also added thought starters to get you moving.

We strongly recommend that you write down this story and keep it. As you move on your journey, take it out, reflect, and add. Your story is a living thing that can serve as a great place for you to document where you have been and where you are going.

1. Who and what were some of the key influences that made you who you are today?

   You are not a lone voice crying in the wilderness. Your life and path intertwined with the paths of others along the way. Who were these people? What did they teach you?

2. What did you dream about becoming when you were growing up?

   Recapture your inner childhood. What excites you? Where do you spend your time?

3. Was there a pivotal moment that set you on your course?

   What dream are you pursuing now? From the universe of potential dreams, why did you choose this one?

4. Do you have a mission in life? How would you describe it?

   Do you want to change the world or just change your own course? What do you know that others need to know? What do you want to discover that others will want to know?

5. What do you think is the difference between people who make an impact on the world and those who do not?

   Not everyone is a Firestarter. Why not? What differentiates those who live their dreams from those whose dreams will never live?

6. Innovators create things. Instigators change things. Initiators start things. How would you identify yourself? Why?

   You may find that different situations require you to adopt different identities. Think of a time when you adopted each of these roles. Did the situation seem natural? Which identity makes you feel most comfortable?

7. What ignites you as a Firestarter?

   Do you search for ways to feel free, unconstrained by external controls in the moment? Do you have a talent that screams to be shared? What passion drives you? In what domain do you feel you are most likely to succeed? Are you true to yourself—do you judge yourself fairly and consistently?

8. What fuels the fire in you?

   In what ways are you powerful? What opportunities have knocked on your door? Did you ignore them or embrace them? How will you find the money needed to finance your journey? Do you pay attention to risk and plan for ways to become lucky? Whom do you know whose energy, resources, and talents you can tap?

9. What has accelerated your success?

   Are your actions consistent with a mission? Do you have a partner with whom you work toward success jointly? What standards do you aim to surpass? Do you work harder than anyone else? How do you identify the teachers and mentors who will help you succeed?

10. What threatens to extinguish your fire, and how can you overcome these Extinguishers?

    Do you have limited resources? Are you afraid, scared, or worried that you will disappoint others? Is there a threat of backlash if you act? Do certain aspects of your life seem uncontrollable? Do you need knowledge or a certain type of skill that you lack? In what ways can you plan around these threats?

Chapter 35

THE FIRESTARTER POTENTIAL SCALE

T he following high-level assessment is a tool to help you look at the impact you might have in a situation. It should be viewed as informational, not determinative. Use this assessment to determine if there are actions you can take to increase the probability of making your impact. Please follow the seven steps below to calculate and assess your Firestarter potential.

IGNITION POTENTIAL

Please rate each item using the following scale:

> 1 = Not at all
> 2 = A little
> 3 = A moderate amount
> 4 = A good amount
> 5 = A lot

Calculation. Add the scores across all rows to calculate the overall Ignition Potential score. This value will be between 5 and 25.

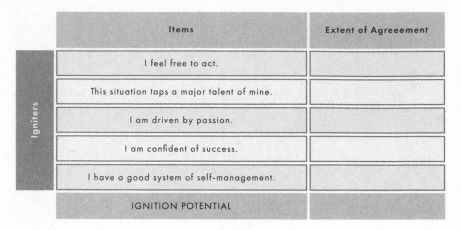

	Items	Extent of Agreeement
Igniters	I feel free to act.	
	This situation taps a major talent of mine.	
	I am driven by passion.	
	I am confident of success.	
	I have a good system of self-management.	
	IGNITION POTENTIAL	

Fig. 35.1. Ignition potential.

FUEL POTENTIAL

In the following table, please rate each item using the following three scales:

Availability. To what extent is the Fuel available in the situation (regardless of how much exists)?

1 = Not at all
2 = A little
3 = A moderate amount
4 = A good amount
5 = A lot

Effective Use. To what extent do you feel you can effectively use the Fuel in the situation?

1 = Not at all
2 = A little
3 = A moderate amount
4 = A good amount
5 = A lot

Calculation. Add the Availability and Degree of Use scores and place the resulting score in the Total column.

Add the total scores across all rows to calculate the GRAND TOTAL. Then divide this score in half to calculate the Fuel Potential score. This value will be between 5 and 25.

	Items	Availability	Effective Use	Total
Fuels	Prestige			
	Opportunity			
	Wealth			
	Luck			
	Social Connections			
	GRAND TOTAL			
	FUEL POTENTIAL (Divide Total in Half)			

Fig. 35.2. Fuel potential.

DISCOURAGERS

In the following table, please rate each item using the following scale:

Degree of Threat. To what extent do the following aspects of the situation threaten your progress?

1 = Not at all
2 = A little
3 = A moderate amount
4 = A good amount
5 = A lot

Calculation. Add the scores across all rows to calculate the overall Discouragers score. This value will be between 5 and 25.

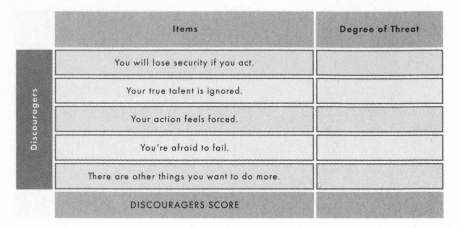

Items	Degree of Threat
You will lose security if you act.	
Your true talent is ignored.	
Your action feels forced.	
You're afraid to fail.	
There are other things you want to do more.	
DISCOURAGERS SCORE	

Fig. 35.3. Discouragers.

FUEL LIMITS

In the following table, please rate each element using the following scale:

> **Limits.** To what extent is the Fuel limited in the situation?
> 1 = Not at all
> 2 = A little
> 3 = A moderate amount
> 4 = A good amount
> 5 = A lot

> **Calculation.** Add scores across all rows to calculate the overall Fuel Limits score. This value will be between 5 and 25.

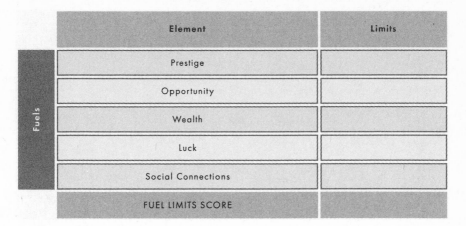

Fig. 35.4. Fuel limits.

ACCELERATION

In the following table, please rate each item using the following scale:

Agreement. To what extent do you agree with the statement about the situation?

1 = Not at all
2 = A little
3 = A moderate amount
4 = A good amount
5 = A lot

Calculation. Add the scores across all rows to calculate the overall Acceleration score. This value will be between 5 and 25.

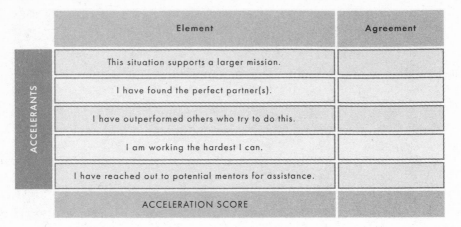

Fig. 35.5. Acceleration.

FIRESTARTER POTENTIAL

Ignition Sub-Total Calculation:

Subtract your Discouragers score from your Ignition Potential. For the Ignition Sub-Total score, write "0" if the result is negative. Otherwise, write the resulting score. This value will be between 0 and 20.

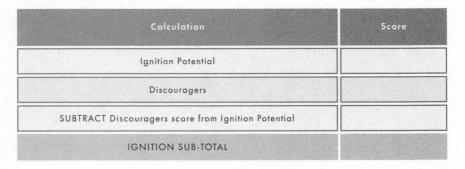

Fig. 35.6. Ignition sub-total calculation.

Fuel Sub-Total Calculation:

Subtract your Fuel Limits score from your Fuel Potential. For the Fuel Sub-Total score, write "0" if the result is negative. Otherwise, write the resulting score. This value will be between 0 and 20.

Calculation	Score
Fuel Potential	
Fuel Limits	
SUBTRACT Fuel Limits score from Fuel Potential	
FUEL SUB-TOTAL	

Fig. 35.7. Fuel sub-total calculation.

Firestarter Potential Calculation:

Calculate the SUM of your Ignition Sub-Total and Fuel Sub-Total. This should provide a number between 0 and 40. Multiply the SUM by your Acceleration score. This Firestarter Potential value will be between 0 and 1,000.

Calculation	Score
Ignition Sub-Total	
Fuel Sub-Total	
SUM of Fuel and Ignition Sub-Totals	
Acceleration	
FIRESTARTER POTENTIAL	

Fig. 35.8. Firestarter potential calculation.

To learn more about your score refer to the following key:

Score Range	Potential	Interpretation
401 to 1,000	Great Potential	Why are you wasting your time reading this? You've got to change the world!
101 to 400	Good Potential	You are on to something here!
50 to 100	Moderate Potential	If you act consistently and the conditions are right, you may ignite.
1 to 50	Limited Potential	Pursuing this course of action is extremely risky.
0	No Potential	You should consider another venue for making your impact.

Fig. 35.9. Firestarter potential key.

HOW TO INTERPRET YOUR SCORE

Is your Firestarter potential where you expected it to be? A little higher? A little lower? Let's learn from it.

If your score is what you expected, do the following:

- Look over your answers to each section. Does anything surprise you?
- If you scored lower or higher than anticipated on any item, think it through. Determine if there is an action you can take to make a difference.
- If you have high Firestarter potential (and you anticipated this), determine the best way to bring this potential to fruition.
- If you have low Firestarter potential (and you anticipated this), think about other situations. Take the test again and see if you can find the best situation to maximize your Firestarter potential.

If your score is lower than expected, do the following:

- Retake the survey.
- This time put yourself in the following mindset: what is the "best case scenario" for each score?

- Compare your results for the first survey with the second.
- Determine how you can make your "best case scenario" a reality.

If your score is higher than expected, do the following:

- Make sure your ratings were completely honest.
- Determine the best way to move forward and make your impact.

Chapter 36

ARE YOU AN INNOVATOR, INSTIGATOR, OR INITIATOR?

Look at each pair of statements. Choose the one statement in the pair with which you most identify. If the decision is difficult, choose the one that is either most true or least false.

Write a "1" or "2" in the right-hand column:

	Statement 1	Statement 2	Choice (1 or 2)
1	In group projects, I take the lead.	Other people offer great ideas in group projects.	
2	A small tweak to a product can make a huge difference.	I'd rather play video games than design them.	
3	I plan to ensure a secure future.	There are always new things to discover.	
4	Some people argue for the sake of arguing.	I don't care about building a better mousetrap. How about mouse cars?	
5	I like a project with a measured pace.	I often get anxious about how slowly others move.	
6	I like to see proof before offering my support for a new process.	I would support a new process that could improve results.	
7	The world is so inefficient.	I'm amazed by the progress of technology.	
8	It's okay to propose multiple wrong ideas to find the right one.	It's better to think things through before presenting ideas.	
9	I like to sit back and observe what other people do.	It's always best to make the first move.	
10	My ideas are usually better than others.	I defer to those who know more than I do.	
11	I see the possibility in any situation.	I assess risks and plan for them.	

	Statement 1	Statement 2	Choice (1 or 2)
12	I offer sound, logical options.	People marvel at my creative ideas.	
13	It's always important to have a well laid-out plan.	Sometimes you have to move forward without knowing your destination.	
14	I look forward to the start of a new day.	I just want the world to slow down sometimes.	
15	Problems usually have one best answer.	I can make this true: 2 + 2 = 5.	
16	Everything you believe could be wrong.	It's best to pick a direction and stay the course.	
17	I like to master each step of a process in sequence.	I get excited when new ideas work as planned.	
18	When you take the lead, you inspire others.	It's important for the most qualified person to take the lead.	
19	Learning never stops.	At some point, it's more beneficial to teach others to learn new things.	
20	I like to pick the perfect outfit to make an impression.	I want to change the world.	
21	I love a relaxing vacation.	I never stop moving.	

Fig. 36.1. Innovator, Instigator, or Initiator.

SCORING THE ASSESSMENT

A score of 5 or higher on any scale indicates you have the potential to tap that Firestarter identity. Scores lower than 5 on any scale indicate it may be more difficult for you to approach situations from the perspective of a particular identity.

Look for these responses:		How many of your responses matched?	Score for:	Is it 5 or more?
Q1: 1 Q5: 2 Q9: 2	Q13: 1 Q14: 1 Q18: 1 Q21: 2		Initiator	
Q2: 1 Q3: 2 Q6: 2	Q11: 1 Q12: 2 Q17: 2 Q19: 1		Innovator	
Q4: 2 Q7: 1 Q8: 1	Q10: 1 Q15: 2 Q16: 1 Q20: 2		Instigator	

Fig. 36.2. Scoring Innovator, Instigator, or Initiator.

Chapter 37

ARE YOU A SUPPORTER OR
AN EXTINGUISHER?

Think about a person in your life who has great aspirations and Firestarter potential. Answer the following questions to determine if you are a Supporter or an Extinguisher of their fire. And, yes, you can be both. There are five steps to this exercise.

Look at each pair of statements. Choose the one statement in the pair with which you most identify. If the decision is difficult, choose the one that is either most true or least false.

Write a "1" or "2" in the right-hand column:

	Statement 1	Statement 2	Choice
1	I pay close attention to the details of his/her plans.	I ask him/her questions to find out more about his/her personality.	
2	I offer him/her advice about life.	I make his/her ideas better.	
3	I do NOT agree with how the person spends money.	I would NOT spend money to support the person if he/she needed it.	
4	I have interesting conversations with the person.	I tell the person about trends that I see in the world.	
5	I make sure he/she feels good about his/her ideas.	I make sure he/she is in a good mood.	
6	I warn the person about the obstacles in his/her way.	I worry that the person will fail.	
7	Sometimes, I don't think the person knows what he/she is doing.	I don't intervene even when I know he/she will fail.	
8	I let the person know when he/she is going down the wrong path.	I let the person know if I change my mind about something we discussed.	
9	I do nice things for the person.	I listen to his/her ideas.	
10	I am often busy when he/she wants to talk.	I have more things to get done than he/she knows about.	
11	I compliment the person's good ideas.	I find ways to apply the person's ideas to my life.	
12	I avoid the person when he/she is having a bad day.	I avoid the person when he/she is being self-destructive.	

	Statement 1	Statement 2	Choice
13	I find ways to make the person feel special.	I find ways to let the person know I understand his/her drive.	
14	I often ignore the person.	The person sometimes bores me.	
15	I don't talk to the person about his/her failures.	I don't talk to the person about controversial topics.	
16	I would help the person through a difficult time.	It makes me feel good to help him/her.	
17	I don't like some of the person's habits.	I let the person know when he/she annoys me.	
18	I would help to grow his/her ideas.	I would help him/her discover new hobbies.	
19	I prevent a bad mood from affecting him/her.	I stop others from criticizing him/her.	
20	I can't believe how stupid he/she acts sometimes.	I tease the person when he/she does dumb things.	
21	I cringe at some of the decisions he/she makes.	I steer the person away from risky actions.	
22	I get mad at the person for making a poor decision.	I get upset if the person ignores my advice.	
23	I encourage the person to pursue his/her passions.	I admire the person's passions.	
24	I find ways to make the person happy.	I find ways to keep the person motivated.	

Fig. 37.1. Supporter or Extinguisher.

SCORE THE ASSESSMENT

A score of 2 or higher on any scale indicates you are tapping the associated Supporter or Extinguisher role.

Look for these responses:	How many of your responses matched?	Score for:	Is it 2 or more?
Q1: 1 Q9: 2 Q13: 2		Nurturer (Supporter)	
Q11: 1 Q23: 1 Q24: 2		Motivator (Supporter)	
Q2: 2 Q4: 2 Q18: 1		Illuminator (Supporter)	
Q5: 1 Q16: 1 Q19: 2		Protector (Supporter)	
Q3: 2 Q10: 1 Q14: 1		Limiter (Extinguisher)	
Q6: 1 Q8: 1 Q21: 2		Discourager (Extinguisher)	
Q17: 2 Q20: 2 Q22: 1		Punisher (Extinguisher)	
Q7: 2 Q12: 2 Q15: 1		Enabler (Extinguisher)	

Fig. 37.2. Scoring Supporter or Extinguisher.

INTERPRET YOUR SCORE

Are you supporting or extinguishing the other person in ways that you anticipated? If not, the good news is that you can take action to change.

To be a more effective Supporter, you can channel your energy in the following ways:

- Convert your positive energy into action. Thinking great things about someone else is empowering in itself, but try to translate your thoughts into behaviors that help the other person grow.

- Focus on the other person. Look at your answers to the questions. Did many of your responses focus on how the other person made *you* feel? Switch that narrative and find ways that you can bolster the other person's confidence.
- Enhance what makes the person a Firestarter. You may have great interactions, but your conversations and actions may be focused on aspects of the person's life that have nothing to do with his or her Firestarter potential. To be a better Supporter, find ways directly to help with that person's impact, rather than distracting the person from it.

To reduce your Extinguisher tendencies, you can edit your behavior in the following ways:

- Reflect on your negative energy. Determine if the way you're treating the other person is a projection of your feelings or an objective review of their actions. Take time to cool off before reacting to behaviors that disappoint you.
- Focus on the big things. Does every nitpick you point out really matter? What seems like something small to you may end up as a bigger issue in the long run, especially if you continually point out similar small things. Before you say something or try to direct the other person's actions, ask yourself if you are preventing something negative or just inserting your preferences into his or her life.
- Recognize patterns. Help the person learn from mistakes—not by relentlessly pointing them out. Rather, guide the person in developing a more self-reflective attitude so that he or she can discover some of the solutions to his or her own recurring issues.

Chapter 38

ASSESSING YOUR SOCIAL BALANCE

In the chapter about Accelerants, we discussed how actions that spread your fire could be described in terms of their alignment with a Socially Balanced Strategy. This exercise allows you to assess your own social balance in the approach you choose to adopt in spreading your fire.

INSTRUCTIONS

Think about an area of your life where you desire to be a Firestarter. For each statement below, shade the boxes based on the following key:

1. Leave the box alone if the statement does not apply.
2. Shade half the box if the statement applies but you are inconsistent in putting it into action.
3. Shade the full box if the statement applies and you are consistently effective in putting it into action.

Self-Improvement		Making an Impact on Others	
Mission Focus: You have a mission that results in treating yourself right.		Mission Focus: You have a mission that results in making an impact on others.	
Collaboration: You bring multiple talents to bear and are able to combine them in unique ways with others to improve your outcomes.		Collaboration: You find ways to improve joint outcomes with others through partnerships.	
Sweat Equity: You find personal joy in working harder than anyone else.		Sweat Equity: Your potential to make an impact drives you to overcome challenges.	
Constructive Competition: You compete with yourself to surpass your expectations.		Constructive Competition: You compete with others to make the biggest impact.	
Support Seeking: You seek mentors and supporters to help you improve your knowledge and skills.		Support Seeking: You seek mentors and supporters to help you make the greatest impact.	

Fig. 38.1. Social balance.

LOOKING FOR PATTERNS

After you complete the exercise, search your results for patterns. Are there too many boxes with no shading or half shading? If so, is the absence of consistency something you accept or something you're surprised about? Does one column have more shading than the other? Is your focus mainly on yourself or mainly on the impact you're trying to make? Does the result surprise you?

There is no right or wrong pattern!

While having more fully shaded boxes is more likely to accelerate your fire, some people target their energy for good reasons. You can't do everything all the time. You have to prioritize. The important thing is for the pattern of your actions to be purposeful. Think carefully about each box. Determine if this is how you want to present yourself. If not, use the following chart to record how you would like to see changes in your pattern to become more socially balanced.

Accelerant	Plan of Action
Mission Focus	
Collaboration	
Sweat Equity	
Constructive Competition	
Support Seeking	

Fig. 38.2. Accelerant plan of action.

Chapter 39

HOW KEY INFLUENCES CONVERGE

We have described cognitive convergence as the phenomenon that allows Firestarters to adapt easily to new contexts. Firestarters apply lessons they've learned to new situations and experience success because they replicate the mindset that drives their achievement.

In the following exercise, you will think about both the people who and the experiences that have shaped you. Through reflection, you will develop a *life map* that shows how your key influences converge across your past, present, and future.

To aid in this exercise, we ask you also to reflect on the types of Supporters we have discussed. For ease of recall, they are the following:

1. Nurturers: people who listen to Firestarters and allow ideas to flourish.
2. Motivators: people who focus on others' passions and work to grow their influence.
3. Illuminators: people who shed light on hard-to-grasp concepts to improve a Firestarter's ideas.
4. Protectors: people who accept a Firestarter's eccentricities and defend their path against those who don't understand.

In the chart below, think about your past, present, and future. For the past, identify the people who played each of these roles in your journey and the lessons you learned from them. For the present, ask yourself who is playing these roles now. For the future, ask yourself who you need to play these roles and what you need that person to do.

	Supporter Type	Who	What did (will) you learn?
Past	Nurturer		
	Motivator		
	Illuminator		
	Protector		
Present	Nurturer		
	Motivator		
	Illuminator		
	Protector		
Future	Nurturer		
	Motivator		
	Illuminator		
	Protector		

Fig. 39.1. Past, present, and future Supporters.

In the next chart, think about the Igniters that you have deployed. For the past, identify the strengths you displayed and weaknesses you overcame in putting the Igniters into action. For the present, ask yourself what strengths and challenges you're facing now in igniting. For the future, ask yourself what strengths and challenges you anticipate for each Igniter.

Now compare your two charts. Together they form a life map, combining the people who and the experiences that have paved your path. Think intently about how the people you listed drove or were driven by the experiences you described. Think about the lessons that you learned that helped you capitalize on your strengths and the lessons you wish you had learned that would have allowed you to overcome your weaknesses.

This is your life. This is your map. Embrace it and ignite.

	Igniter	Strengths	Weaknesses & Challenges
Past	Freedom in the Moment		
	Tapped Talent		
	Mastery Mindset		
	Driving Passion		
	Global Introspection		
Present	Freedom in the Moment		
	Tapped Talent		
	Mastery Mindset		
	Driving Passion		
	Global Introspection		
Future	Freedom in the Moment		
	Tapped Talent		
	Mastery Mindset		
	Driving Passion		
	Global Introspection		

Fig. 39.2. Past, present, and future Igniters.

Chapter 40

FOUR-DIMENSIONAL GOALS TEMPLATE

I n our discussion of global introspection, we described how Firestarters view goals in four dimensions. By doing this, they thwart the feelings of failure that arise when specific targets aren't reached. This brief template will help you develop your own four-dimensional goals.

In the example shown, the fictional person failed to achieve the goal in two of the dimensions. But examine the brief descriptions in the Lessons Learned and Interpretation column. Can you see how reviewing goals and results in this four-dimensional approach would thwart feelings of failure?

Dimension	Goal	Result	Lessons Learned and Interpretation
1: Standard based on expected performance.	Describe overall performance standard based on past performance or overall expectation. Exceed best time of seven minutes for running a mile during the season.	Record actual result. Best time: 7 minutes 25 seconds.	Discuss and analyze results of meeting or failing to meet target. My training regiment was disrupted due to a sprained ankle.
2: Standard based on situational constraints.	Describe performance standard based on known constraints. Taking into account my recent sprained ankle, come within 30 seconds of my best time (7:00) this season.	Record actual result. Won two races despite not reaching best time.	Discuss and analyze results of meeting or failing to meet target. Great performance. I pushed myself hardest despite physical constraints.

Dimension	Goal	Result	Lessons Learned and Interpretation
3: Standard based on others' expectations.	Describe performance standard that others expect. Coach expects me to try my hardest and hopefully be at 6:30 by the end of the season.	Record actual result. Best time: 7 minutes 25 seconds.	Discuss and analyze results of meeting or failing to meet target. Coach was proud of the heart I displayed and my effort throughout the season.
4: Standard based on out-performing others.	Describe performance standard based on competitive realities. Win at least 1 race, regardless of times achieved.	Record actual result. Won 2 races despite not reaching best time.	Discuss and analyze results of meeting or failing to meet target. I was victorious. I paced myself due to prior injury but managed to put in the right amount of effort to take the gold twice.

Fig. 40.1. Four-dimensional goals.

LIGHT THE MATCH!

Our mission is to ignite your Firestarter potential. We also hope to expand the thinking of those who are already making an impact. We have described leading research, shared anecdotes of some intriguing Firestarters throughout history, and provided profiles of leading Firestarters today. We have developed tools and assessments to guide your journey, and you've learned how to get ignited, identify Fuels, utilize Accelerants, and avoid Extinguishers to reach your Firestarter potential. The Firestarter Framework speaks to situations you may face on the road to finding your inner Firestarter. There is only one thing left for you to do at this point. Light the match!

We would like to continue down the Firestarter road with you. We invite you to visit our online portal at www.gofirestarter.com. You'll gain access to exclusive content such as in-depth profiles and exercises. You'll find out about coaching, workshops, speaking engagements, and other Firestarter services to help ignite, fuel, and accelerate yourself or your organization. Light the match!

We don't know how your ideas will evolve. We do know that if you don't pursue them, the world may not be as bright. Tap your talent and passion, look to the Firestarters in this book as inspiration, and always remember that you have the power to generate Emissions that will improve your children's lives, future generations, and in some way shape the world. Figure out what burns in you. Light the match!

NOTES

CHAPTER 1. WHY WE BECAME CURIOUS ABOUT FIRESTARTERS

1. "Mars: Novo Mundo," season 1, episode 1, directed by Everardo Gout, aired November 14, 2016 (Erfoud, Morocco: National Geographic, 2016), http://channel .nationalgeographic.com/mars/episodes/novo-mundo/ (accessed June 29, 2017).

2. Deborah Castellano Lubov, "It's Official: 'Saint' Mother Teresa," *Zenit*, September 4, 2016, https://zenit.org/articles/its-official-saint-mother-teresa (accessed June 30, 2017).

3. "John Adams," Biography.com, last updated April 27, 2017, https://www .biography.com/people/john-adams-37967 (accessed June 30, 2017).

4. Ibid.

CHAPTER 2. THE FIRESTARTER FRAMEWORK

1. Albert Bandura, "Social Cognitive Theory of Self-Regulation," *Organizational Behavior and Human Decision Processes* 50, no. 2 (1991): 248–87.

2. Melissa S. Cardon et al., "The Nature and Experience of Entrepreneurial Passion," *Academy of Management Review* 34, no. 3 (2009): 511–32.

3. Sheldon Stryker and Peter J. Burke, "The Past, Present, and Future of an Identity Theory," *Social Psychology Quarterly* (2000): 284–97.

4. Matthew Toren, "6 Genuine Reasons Why People Become Entrepreneurs," Entrepreneur.com, October 22, 2015, https://www.entrepreneur.com/article/251838 (accessed June 16, 2017).

5. Management Innovation Exchange, "Gary Hamel," http://www.management exchange.com/users/ghamel (accessed September 8, 2017).

6. Fiona Patterson and Maire Kerrin, "Great Minds Don't Think Alike: Person-Level Predictors of Innovation at Work," in *Creativity in Arts, Science, and Technology*, ed. F. K. Reisman (Leeds, UK: Knowledge, Innovation, and Enterprise Conference, 2016), pp. 58–88.

7. Joe Hagan, "Tom Freston, Runaway Mogul," *Men's Journal*, February 13, 2013, http://www.mensjournal.com/magazine/tom-freston-runaway-mogul-20130213 (accessed September 8, 2017).

8. "Ada Lovelace," Biography.com, last updated July 10, 2017, https://www
.biography.com/people/ada-lovelace-20825323 (accessed August 15, 2017).

9. John Sculley, in interview with Raoul Davis, March 29, 2017.

10. Teresa M. Amabile, *Creativity in Context: Update to the Social Psychology of
Creativity* (Boulder: Westview, 1996).

11. Richard W. Woodman, John E. Sawyer, and Ricky W. Griffin, "Toward a Theory
of Organizational Creativity," *Academy of Management Review* 18, no. 2 (1993): 293–321;
Richard W. Woodman, and Lyle F. Schoenfeldt, "An Interactionist Model of Creative
Behavior," *Journal of Creative Behavior* 24, no. 1 (1990): 10–20.

12. Luke Hackett, "Shigeru Miyamoto Biography: His Early Life & Career to
Modern Day," Super Luigi Bros, 2017, http://www.superluigibros.com/shigeru
-miyamoto-biography (accessed June 16, 2017).

13. "Christopher Columbus," Biography.com, last updated August 1, 2017, https://
www.biography.com/people/christopher-columbus-9254209 (accessed August 16, 2017).

14. Tupac Shakur, "Hail Mary," written by Tupac Shakur, Young Noble, and Kastro,
on *The Don Killuminati: The 7 Day Theory*, released November 5, 1996, Death Row/
Interscope.

15. "Clara Barton," Biography.com, last updated April 27, 2017, https://www
.biography.com/people/clara-barton-9200960 (accessed June 16, 2017).

16. Paul B. Brown, "What Makes Someone an Entrepreneur?" *Forbes*, May 6, 2013,
https://www.forbes.com/sites/actiontrumpseverything/2013/05/06/what-makes
-someone-an-entreprenur-7/#127baaab4ab2 (accessed June 16, 2017).

17. Mark H. Davis, Jennifer A. Hall, and Pamela S. Mayer, "Developing a New
Measure of Entrepreneurial Mindset: Reliability, Validity, and Implications for
Practitioners," *Consulting Psychology Journal: Practice and Research* 68, no. 1 (2016): 1–28.

18. For example, see Richard W. Woodman, John E. Sawyer, and Ricky W. Griffin,
"Toward a Theory of Organizational Creativity," *Academy of Management Review* 18, no. 2
(1993): 293–321.

19. Professional trainer, in personal conversation with Raoul Davis, May 14, 2016.

20. Tyler Kepner, "Kobe Bryant Ends Career with Exclamation Point, Scoring 60
Points," *New York Times*, April 14, 2016, https://www.nytimes.com/2016/04/15/sports/
basketball/kobe-bryant-scores-60-for-los-angeles-lakers-in-farewell-game.html?mcubz=0
(accessed September 8, 2017).

21. "Oprah Winfrey," Biography.com, April 27, 2017, https://www.biography.com/
people/oprah-winfrey-9534419 (accessed June 16, 2017).

22. Ibid.

23. Daniel Rapaport, "The Biggest Comebacks in Super Bowl History," *Sports
Illustrated*, February 5, 2017, https://www.si.com/nfl/2017/02/05/super-bowl-biggest
-comebacks-patriots (accessed September 8, 2017).

24. Robert Eisenberger and Judy Cameron, "Detrimental Effects of Reward: Reality
or Myth?" *American Psychologist* 51, no. 11 (1996): 1153–66.

25. Edward L. Deci and Richard M. Ryan, *Intrinsic Motivation and Self-Determination in Human Behavior* (New York: Plenum, 1985).

26. Robert Eisenberger and Linda Shanock, "Rewards, Intrinsic Motivation, and Creativity: A Case Study of Conceptual and Methodological Isolation," *Creativity Research Journal* 15, no. 2-3 (2003): 121–30.

27. Peter J. Burke and Donald C. Reitzes, "The Link between Identity and Role Performance," *Social Psychology Quarterly* (1981): 83–92.

28. Robert Eisenberger, "Learned Industriousness," *Psychological Review* 99, no. 2 (1992): 248–67.

29. Steven Marcus, "Jackie Robinson and Branch Rickey: Together in History," *Newsday*, February 25, 2017, http://www.newsday.com/sports/baseball/jackie-robinson -and-branch-rickey-together-in-history-1.13174735 (accessed June 17, 2017).

30. Sam Costello, "How Many Apps Are in the App Store?" Lifewire, May 5, 2017, https://www.lifewire.com/how-many-apps-in-app-store-2000252 (accessed June 17, 2017).

31. Harry Pettit, "The Billionaire So Tired of Being Stuck in LA Traffic He Is Creating a TUNNEL to Get to Work: Elon Musk Reveals First Image of His Underground Solution to Congestion," *Daily Mail*, February 6, 2017, http://www .dailymail.co.uk/sciencetech/article-4195704/Elon-Musk-tweets-photo-giant-tunnel -boring-machine.html (accessed June 17, 2017).

32. Darren Orf, "10 Awesome Accidental Discoveries," *Popular Mechanics*, June 27, 2013, http://www.popularmechanics.com/science/health/g1216/10-awesome-accidental -discoveries/ (accessed June 17, 2017).

CHAPTER 3. IGNITION POTENTIAL

1. Daniel H. Pink, *Drive: The Surprising Truth about What Motivates Us* (New York: Penguin, 2011).

2. J. Robert Baum and Edwin A. Locke, "The Relationship of Entrepreneurial Traits, Skill, and Motivation to Subsequent Venture Growth," *Journal of Applied Psychology* 89, no. 4 (2004): 587–98.

CHAPTER 4. FREEDOM IN THE MOMENT: LIVE ON YOUR TERMS

1. Don Miguel Ruiz Jr., in interview with Kathy Palokoff, January 17, 2017.

2. Julian B. Rotter, "Internal Versus External Control of Reinforcement: A Case History of a Variable," *American Psychologist* 45, no. 4 (1990): 489–93.

3. Edward L. Deci and Richard M. Ryan, "The 'What' and 'Why' of Goal Pursuits:

Human Needs and the Self-Determination of Behavior," *Psychological Inquiry* 11, no. 4 (2000): 227–68.

    4. Ibid.

CHAPTER 5. TAPPED TALENT: USE THE GIFTS YOU HAVE

    1. "Wolfgang Mozart," Biography.com, last updated April 27, 2017, https://www.biography.com/people/wolfgang-mozart-9417115 (accessed June 17, 2017).

    2. Scott Adams, *How to Fail at Almost Everything and Still Win Big: Kind of the Story of My Life* (New York, Penguin, 2013).

    3. Edward L. Deci and Richard M. Ryan, *Intrinsic Motivation and Self-Determination in Human Behavior* (New York: Plenum, 1985).

    4. Edward L. Deci, Richard Koestner, and Richard M. Ryan. "A Meta-Analytic Review of Experiments Examining the Effects of Extrinsic Rewards on Intrinsic Motivation," *Psychological Bulletin* 125, no. 6 (1999): 627–68.

    5. Ibid.

    6. J. Robert Baum and Edwin A. Locke, "The Relationship of Entrepreneurial Traits, Skill, and Motivation to Subsequent Venture Growth," *Journal of Applied Psychology* 89, no. 4 (2004): 587–98.

CHAPTER 6. MASTERY MINDSET: BE CONFIDENT IN WHO YOU ARE

    1. Barack Obama, *The Audacity of Hope: Thoughts on Reclaiming the American Dream* (New York: Crown, 2006).

    2. Albert Bandura, "Self-Efficacy: Toward a Unifying Theory of Behavioral Change," *Psychological Review* 84, no. 2 (1977): 191–215.

    3. Albert Bandura, *Social Foundation of Thought and Action: A Social Cognitive Theory* (Prentice Hall: New York, 1986).

    4. Albert Bandura, "Social Cognitive Theory of Self-Regulation," *Organizational Behavior and Human Decision Processes* 50, no. 2 (1991): 248–87.

    5. Eric T. Wagner, "Five Reasons 8 out of 10 Businesses Fail," *Forbes*, September 12, 2013, https://www.forbes.com/sites/ericwagner/2013/09/12/five-reasons-8-out-of-10-businesses-fail (accessed June 17, 2017).

    6. Bandura, "Social Cognitive Theory."

    7. Albert Bandura and Daniel Cervone, "Differential Engagement of Self-Reactive Influences in Cognitive Motivation," *Organizational Behavior and Human Decision Processes* 38, no. 1 (1986): 92–113.

    8. Bandura, "Social Cognitive Theory."

9. Bandura, "Self-Efficacy."

10. Ibid.

11. Ibid.

CHAPTER 7. DRIVING PASSION: DO WHAT YOU LOVE

1. Mary Dejevsky, "The First Woman in Space: 'People Shouldn't Waste Money on Wars,'" *Guardian*, March 29, 2017, https://www.theguardian.com/global-development -professionals-network/2017/mar/29/valentina-tereshkova-first-woman-in-space-people -waste-money-on-wars (accessed June 17, 2017).

2. Marie Claire, "These Are the Most Inspirational Women in History," *Marie Claire*, March 21, 2017, http://www.marieclaire.co.uk/entertainment/people/inspirational -women-from-history-81054 (accessed June 17, 2017).

3. W. David Pierce et al., "Positive Effects of Rewards and Performance Standards on Intrinsic Motivation," *Psychological Record* 53, no. 4 (2003): 561–79.

4. Mihaly Csikszentmihalyi, *Flow: The Psychology of Optimal Performance* (New York: Cambridge University Press, 1990).

5. Minet Schindehutte, Michael Morris, and Jeffrey Allen, "Beyond Achievement: Entrepreneurship as Extreme Experience," *Small Business Economics* 27, no. 4 (2006): 349–68.

6. Melissa S. Cardon et al., "The Nature and Experience of Entrepreneurial Passion," *Academy of Management Review* 34, no. 3 (2009): 511–32.

7. Ibid.

8. Edwin A. Locke, *The Prime Movers* (New York: Amacom, 2000).

9. Melissa S. Cardon et al., "The Nature and Experience of Entrepreneurial Passion," *Academy of Management Review* 34, no. 3 (2009): 511–32.

10. Edward L. Deci and Richard M. Ryan, "The '"What"' and '"Why"' of Goal Pursuits: Human Needs and the Self-Determination of Behavior," *Psychological Inquiry* 11, no. 4 (2000): 227–68.

11. Christina E. Shalley, Jing Zhou, and Greg R. Oldham, "The Effects of Personal and Contextual Characteristics on Creativity: Where Should We Go from Here?" *Journal of Management* 30, no. 6 (2004): 933–58.

12. Leonard L. Martin et al., "Mood as Input: People Have to Interpret the Motivational Implications of Their Moods," *Journal of Personality and Social Psychology* 64, no. 3 (1993): 317–26.

13. Edward L. Deci, Richard Koestner, and Richard M. Ryan, "A Meta-Analytic Review of Experiments Examining the Effects of Extrinsic Rewards on Intrinsic Motivation," *Psychological Bulletin* 125, no. 6 (1999): 627–68.

14. Ibid.

CHAPTER 8. GLOBAL INTROSPECTION: CONQUER THE SELF

1. Mike Larkin, "What Will Angelina Say? Brad Pitt Has Some Explaining to Do after Buying Soviet Era Tank," *Daily Mail*, November 12, 2011, http://www.dailymail.co.uk/tvshowbiz/article-2060745/What-Angelina-say-Brad-Pitt-explaining-buying-Soviet-era-tank.html (accessed June 17, 2017).

2. Anthony Sulla-Heffinger, "Goldberg on Return to WWE: 'I'm Absolutely Miserable,'" Yahoo! Sports, March 29, 2017, https://sports.yahoo.com/news/goldberg-on-return-to-wwe-im-absolutely-miserable-203527527.html (accessed June 17, 2017).

3. *Hamlet*, ed. Barbara A. Mowat and Paul Werstine, updated ed. (New York: Simon and Schuster Paperbacks, 2012), 1.3.84. References are to act, scene, and line.

4. Albert Bandura, "Social Cognitive Theory of Self-Regulation," *Organizational Behavior and Human Decision Processes* 50, no. 2 (1991): 248–87.

5. Ibid.

6. Brad Gagnon, "Wide Right 25 Years Later: A Super Bowl So Much Larger than Just Scott Norwood," Bleacher Report, January 27, 2016, http://bleacherreport.com/articles/2607652-wide-right-25-years-later-a-super-bowl-so-much-larger-than-just-scott-norwood (accessed June 17, 2017).

7. Sam Galanis, "Adam Vinatieri's Snow Bowl Field Goal Voted Most Memorable Patriots Play," NESN, July 11, 2014, http://nesn.com/2014/07/adam-vinatieris-snow-bowl-field-goal-voted-top-patriots-play (accessed June 17, 2017).

8. Edwin A. Locke and Gary P. Latham, *A Theory of Goal Setting & Task Performance* (Englewood Cliffs, NJ: Prentice-Hall, 1990).

9. Edwin A. Locke, Gary P. Latham, and Miriam Erez, "The Determinants of Goal Commitment," *Academy of Management Review* 13, no. 1 (1988): 23–39.

10. Albert Bandura and Daniel Cervone, "Self-Evaluative and Self-Efficacy Mechanisms Governing the Motivational Effects of Goal Systems," *Journal of Personality and Social Psychology* 45, no. 5 (1983): 1017–28.

11. Albert Bandura, *Social Foundation of Thought and Action: A Social Cognitive Theory* (Prentice Hall: New York, 1986).

12. James J. Gross and Oliver P. John, "Individual Differences in Two Emotion Regulation Processes: Implications for Affect, Relationships, and Well-Being," *Journal of Personality and Social Psychology* 85, no. 2 (2003): 348–62.

13. Ibid.

14. Bandura, "Social Cognitive Theory."

15. John M. Gottman and Richard M. McFall, "Self-Monitoring Effects in a Program for Potential High School Dropouts: A Time-Series Analysis," *Journal of Consulting and Clinical Psychology* 39, no. 2 (1972): 273–81.

CHAPTER 9. FUELING THE FIRE

1. Mark Goulston, "Would You Rather Be a King or Be Rich?" *Business Insider*, August 23, 2010, http://www.businessinsider.com/usable-insight-would-you-rather-be -king-or-rich-2010-8 (accessed June 17, 2017).

2. William D. Bygrave, "Theory Building in the Entrepreneurship Paradigm," *Journal of Business Venturing* 8, no. 3 (1993): 255–80.

3. Jerrie Ueberle, in interview with Kathy Palokoff, February 14, 2017.

CHAPTER 10. PRESTIGE: DELIVER ON YOUR PROMISE

1. John W. Thibaut and Harold H. Kelley, *Interpersonal Relations: A Theory of Interdependence* (New York: John Wiley and Sons, 1978).

CHAPTER 11. OPPORTUNITY: OWN YOUR MOMENT

1. "Kathy Ireland," Biography.com, last updated April 2, 2014, https://www .biography.com/people/kathy-ireland-20978315 (accessed June 17, 2017).

2. Quote attributed to Yogi Berra.

3. Cited in *Wikipedia*, s.v. "Jack Ma," last updated August 16, 2017, https:// en.wikipedia.org/wiki/Jack_Ma (accessed August 16, 2017).

4. Ibid.

5. Steve Pemberton, in personal conversation with Raoul Davis, June 10, 2017.

CHAPTER 12. WEALTH: PAY IT FORWARD

1. Corinne Heller, "*Justice League* Trailer Speaks the Cold, Hard Truth about Batman," E! News, March 25, 2017, http://www.eonline.com/news/838751/justice -league-trailer-speaks-the-cold-hard-truth-about-batman (accessed September 6, 2017).

2. Robert Kiyosaki, "Rich Dad Fundamentals: OPM," RichDad.com, June 7, 2011, http://www.richdad.com/Resources/Rich-Dad-Financial-Education-Blog/June-2011/ Rich-Dad-Fundamentals-OPM.aspx (accessed June 17, 2017).

3. "Stats," Kickstarter, https://www.kickstarter.com/help/stats (accessed September 12, 2017).

4. Krystine Therriault, "Top 10 Most Funded Kickstarter Projects Revisited in 2016," CrowdCrux.com, http://www.crowdcrux.com/top-10-funded-kickstarter-projects -revisited-2016/ (accessed June 17, 2017).

CHAPTER 13. LUCK: PLAY THE ODDS

1. Jacque Wilson, "Viagra: The Little Blue Bill That Could," CNN.com, March 27, 2013, http://www.cnn.com/2013/03/27/health/viagra-anniversary-timeline (accessed September 12, 2017).

2. Ben Mezrich, *Bringing Down the House: The Inside Story of Six M.I.T. Students Who Took Vegas for Millions* (New York: Simon & Schuster, 2002).

CHAPTER 14. SOCIAL CONNECTIONS: INSPIRE LOYALTY

1. Laura Schreffler, "Man of Distinction: Ryan Seacrest Strives for Greatness Every Day," Haute Living, January 15, 2015, http://hauteliving.com/2015/01/man-distinction -ryan-seacrest-strives-greatness-every-day/541680 (accessed May 5, 2017).

2. Patrick Ip, in interview with the author, January 19, 2017.

CHAPTER 15. ACCELERATING YOUR FIRE

1. Paul Eder and Everett Marshall, "Optimizing Results through Socially Balanced Strategies," *OD Practitioner* 42, no. 3 (2010): 30–35.

2. Charles G. McClintock, "Social Values: Their Definition, Measurement and Development," *Journal of Research & Development in Education* (1978): 121–37.

CHAPTER 16. MISSION-FOCUSED BEHAVIOR: TAKE THE LONG ROAD

1. "Roald Amundsen: 1872–1928," South-Pole.com, http://www.south-pole.com/ main.html# (accessed June 24, 2017).

2. TOMS, "Improving Lives," http://www.toms.com/improving-lives (accessed September 12, 2017).

3. Pastor Andy Thompson, in personal conversation with Raoul Davis, November 2, 2016.

4. "Martin Luther King Jr. and the Global Freedom Struggle," KingEncyclopedia. Stanford.edu, http://kingencyclopedia.stanford.edu/encyclopedia/chronologyentry/1965 _03_09.1.html (accessed June 24, 2017).

5. Mihaly Csikszentmihalyi, *Flow: The Psychology of Optimal Performance* (New York: Cambridge University Press, 1990); Jeanne Nakamura and Mihaly Csikszentmihalyi, "The Concept of Flow," in *Flow and the Foundations of Positive Psychology* (Dordrecht: Springer, 2014), pp. 239–63.

6. Csikszentmihalyi, *Flow.*

7. Ibid.

8. J. Robert Baum and Edwin A. Locke, "The Relationship of Entrepreneurial Traits, Skill, and Motivation to Subsequent Venture Growth," *Journal of Applied Psychology* 89, no. 4 (2004): 587–98.

CHAPTER 17. COLLABORATION: INSPIRE EACH OTHER

1. Alexandra Alter, "James Patterson Has a Big Plan for Small Books," *New York Times,* March 21, 2016, https://www.nytimes.com/2016/03/22/business/media/james -patterson-has-a-big-plan-for-small-books.html?_r=0 (accessed June 24, 2017).

2. Alice Vincent, "James Patterson: How the Bestseller Factory Works," *Telegraph,* March 20, 2014, http://www.telegraph.co.uk/culture/books/booknews/10711191/James -Patterson-how-the-bestseller-factory-works.html (accessed June 24, 2017).

3. Alter, "James Patterson Has a Big Plan."

4. Robert E. Ployhart et al., "Human Capital Is Dead; Long Live Human Capital Resources!" *Journal of Management* 40, no. 2 (2014): 371–98.

5. Kyle Pomerleau, "An Overview of Pass-Through Businesses in the United States," *Tax Foundation,* January 21, 2015, https://taxfoundation.org/overview-pass-through -businesses-united-states/ (accessed June 24, 2017).

CHAPTER 18. CONSTRUCTIVE COMPETITION:
RISE TO THE OCCASION

1. Judith M. Harackiewicz, George Manderlink, and Carol Sansone, "Rewarding Pinball Wizardry: Effects of Evaluation and Cue Value on Intrinsic Interest," *Journal of Personality and Social Psychology* 47, no. 2 (1984): 287–300.

2. Jerry Suls and Thomas Ashby Wills, *Social Comparison: Contemporary Theory and Research* (Hillsdale, NJ: Lawrence Erlbaum Associates, 1991).

3. Judy Cameron, "Negative Effects of Reward on Intrinsic Motivation—A Limited Phenomenon: Comment on Deci, Koestner, and Ryan," *Review of Educational Research* 71, no. 1 (2001): 29–42.

CHAPTER 19. SWEAT EQUITY: EXERT EFFORT
TO OVERCOME OBSTACLES

1. Martin Seligman, *Learned Helplessness and Depression in Animals and Men* (Morristown, NJ: General Learning, 1976).

2. Robert Eisenberger, "Learned Industriousness," *Psychological Review* 99, no. 2 (1992): 248–67.

3. Robert Eisenberger, Frances Haskins, and Paul Gambleton, "Promised Reward and Creativity: Effects of Prior Experience," *Journal of Experimental Social Psychology* 35, no. 3 (1999): 308–325.

4. Albert Bandura and Forest J. Jourden, "Self-Regulatory Mechanisms Governing the Impact of Social Comparison on Complex Decision Making," *Journal of Personality and Social Psychology* 60, no. 6 (1991): 941–51.

5. Albert Bandura, "Self-Efficacy: Toward a Unifying Theory of Behavioral Change," *Psychological Review* 84, no. 2 (1977): 191–215.

6. Fiona Patterson and Maire Kerrin, "Great Minds Don't Think Alike: Person-Level Predictors of Innovation at Work," in *Creativity in Arts, Science, and Technology*, ed. F. K. Reisman (Leeds, UK: Knowledge, Innovation, and Enterprise Conference, 2016), pp. 58–88.

CHAPTER 20. SUPPORT SEEKING: HUMBLE YOURSELF

1. Richard M. Ryan and Wendy S. Grolnick, "Origins and Pawns in the Classroom: Self-Report and Projective Assessments of Individual Differences in Children's Perceptions," *Journal of Personality and Social Psychology* 50, no. 3 (1986): 550–58.

2. Robert Eisenberger and Florence Stinglhamber, *Perceived Organizational Support: Fostering Enthusiastic and Productive Employees* (Washington, DC: American Psychological Association, 2011).

3. Dr. Anton Robert Berzins, in interview with Paul Eder, February 8, 2017.

CHAPTER 22. DISCOURAGERS: DON'T LIMIT YOURSELF

1. Nico W. Van Ypern and Mariet Hagedoorn, "Do High Job Demands Increase Intrinsic Motivation or Fatigue or Both? The Role of Job Control and Job Social Support," *Academy of Management Journal* 46, no. 3 (2003): 339–48.

2. Edward L. Deci and Richard M. Ryan, "The 'What' and 'Why' of Goal Pursuits: Human Needs and the Self-Determination of Behavior," *Psychological Inquiry* 11, no. 4 (2000): 227–68.

3. Charlie Jane Anders and Michael Ann Dobbs, "15 Classic Science Fiction and Fantasy Novels That Publishers Rejected," Gizmodo.com, October 19, 2010, http://io9 .gizmodo.com/5668053/15-classic-science-fiction-and-fantasy-novels-that-publishers -rejected (accessed June 17, 2017).

4. Edward L. Deci and Richard M. Ryan, *Intrinsic Motivation and Self-Determination in Human Behavior* (New York: Plenum, 1985).

5. Deci and Ryan, "'What' and 'Why' of Goal Pursuits."

6. Terry R. Lied and Vahé A. Kazandjian, "A Hawthorne Strategy: Implications for Performance Measurement and Improvement," *Clinical Performance and Quality Health Care* 6 (1998): 201–204.

7. Judy Cameron, "Negative Effects of Reward on Intrinsic Motivation—A Limited Phenomenon: Comment on Deci, Koestner, and Ryan," *Review of Educational Research* 71, no. 1 (2001): 29–42.

8. Ibid.

9. Edward L. Deci, Richard Koestner, and Richard M. Ryan, "A Meta-Analytic Review of Experiments Examining the Effects of Extrinsic Rewards on Intrinsic Motivation," *Psychological Bulletin* 125, no. 6 (1999): 627–68.

10. Cameron, "Negative Effects of Reward."

11. Stephane Condon, "Sebelius: 'Hold Me Accountable for the Debacle' of HealthCare.gov," CBS News, October 30, 2013, http://www.cbsnews.com/news/sebelius-hold-me-accountable-for-the-debacle-of-healthcaregov/ (accessed June 17, 2017).

12. Albert Bandura, "Self-Efficacy: Toward a Unifying Theory of Behavioral Change," *Psychological Review* 84, no. 2 (1977): 191–215.

13. Dario Maestripieri, "What Monkeys Can Teach Us about Human Behavior: From Facts to Fiction," *Psychology Today*, March 20, 2012, https://www.psychologytoday .com/blog/games-primates-play/201203/what-monkeys-can-teach-us-about-human -behavior-facts-fiction (accessed June 17, 2017).

14. Ibid.

15. Melissa S. Cardon et al., "The Nature and Experience of Entrepreneurial Passion," *Academy of Management Review* 34, no. 3 (2009): 511–32.

CHAPTER 23. FUEL LIMITS: DON'T RUN OUT OF GAS

1. Eric Rosenbaum, "The Crucial Money Decision Elon Musk Made When He Was Broke," MSN, April 27, 2017, https://www.msn.com/en-us/money/topstories/the-crucial -money-decision-elon-musk-made-when-he-was-broke/ar-BBArmVv (accessed June 17, 2017).

2. Nate Silver, "Dear Media, Stop Freaking out about Donald Trump's Polls," FiveThirtyEight, November 23, 2016, http://fivethirtyeight.com/features/dear-media -stop-freaking-out-about-donald-trumps-polls/ (accessed June 17, 2017).

3. Nate Silver, "Why Republican Voters Decided on Trump," FiveThirtyEight, May 4, 2016, https://fivethirtyeight.com/features/why-republican-voters-decided-on-trump/ (accessed June 17, 2017).

4. Nate Silver, "How I Acted like a Pundit and Screwed up on Donald Trump," FiveThirtyEight, May 18, 2016, https://fivethirtyeight.com/features/how-i-acted-like-a -pundit-and-screwed-up-on-donald-trump/ (accessed June 17, 2017).

5. Gus Lubin, "Professor Who Correctly Predicted 32 Years of Elections Says Trump Will Win—but There Are Caveats," *Business Insider*, October 30, 2016, http://www.businessinsider.com/lichtman-predicts-trump-victory-with-caveats-2016-10 (accessed June 17, 2017).

CHAPTER 24. SELF-MISMANAGEMENT: DON'T BE MISGUIDED

1. Camila Domonoske, "For First Time in 130 Years, More Young Adults Live with Parents than with Partners," NPR, May 24, 2016, http://www.npr.org/sections/thetwo-way/2016/05/24/479327382/for-first-time-in-130-years-more-young-adults-live -with-parents-than-partners (accessed June 17, 2017).

2. Michael B. Kelley and Pamela Engel, "21 Lottery Winners Who Blew It All," *Business Insider*, February 11, 2015, http://www.businessinsider.com/lottery-winners-who -lost-everything-2015-2 (accessed June 17, 2017).

3. Sandra Gonzalez, "CNN Fires Kathy Griffin," CNN, May 31, 2017, http://money.cnn.com/2017/05/31/media/cnn-kathy-griffin/index.html (accessed June 17, 2017).

4. TMZ, "Barron Trump Thought Beheaded Image Was His Dad," TMZ, May 31, 2017, http://www.tmz.com/2017/05/31/barron-trump-thought-donald-beheaded-image -real/ (accessed June 24, 2017).

5. James J. Gross and Oliver P. John, "Individual Differences in Two Emotion Regulation Processes: Implications for Affect, Relationships, and Well-Being," *Journal of Personality and Social Psychology* 85, no. 2 (2003): 348–62.

6. Martin Seligman, *Learned Helplessness and Depression in Animals and Men* (Morristown, NJ: General Learning, 1976).

7. Polly Mosendz and Kim Bhasin, "The Inside Story of How Fyre Festival Went Up in Flames: Sexy Ads with Supermodels? Check. Food, Lodging, and Security? Not So Much," *Bloomberg Pursuits*, May 2, 2017, https://www.bloomberg.com/news/articles/2017-05-02/the-inside-story-of-how-fyre-festival-went-up-in-flames (accessed June 24, 2017).

8. Lisa Ryan, "Fyre Festival Organizers Reportedly Blew Their Early Funding on Models and Yachts," Cut, May 3, 2017, https://www.thecut.com/2017/05/fyre-festival -models-early-funding-vice.html (accessed June 24, 2017).

CHAPTER 25. PUNISHERS: DON'T LET OTHERS HURT YOU

1. Eric Lutz, "Bill O'Reilly Was Taken Down by New York Times Reporter He Threatened in 2015," CNBC, April 20, 2017, http://www.cnbc.com/2017/04/20/bill -oreilly-was-taken-down-by-new-york-times-reporter-he-threatened-in-2015.html (accessed June 17, 2017).

2. Emily Steel and Michael S. Schmidt, "Bill O'Reilly Thrives at Fox News, Even as Harassment Settlements Add Up," *New York Times*, April 1, 2017, https://www.nytimes .com/2017/04/01/business/media/bill-oreilly-sexual-harassment-fox-news.html?mcubz=0 (accessed September 8, 2017).

3. Bill O'Reilly, "Statement of Bill O'Reilly on His Departure from Fox News," BillOReilly.com, April 19, 2017, https://www.billoreilly.com/b/Statement-of-Bill -OReilly-on-his-Departure-from-Fox-News/-717897613864653124.html (accessed June 17, 2017).

CHAPTER 26. INEPTITUDE: DON'T BE STUPID

1. Dave Anderson, "Mookie Wilson's Grounder. Bill Buckner's Legs," *New York Times*, October 25, 1986, http://www.nytimes.com/packages/html/sports/year_in _sports/10.25.html?mcubz=0 (accessed September 17, 2017).

CHAPTER 27. INNOVATORS CREATE THINGS

1. This quote and following quotes and information from John Sculley, in interview with Raoul Davis, March 29, 2017.

2. "John Sculley," Biography.com, last updated April 2, 2014, https://www.biography .com/people/john-sculley-21187457#! (accessed September 18, 2017).

3. This quote and following quotes and information from Maya Penn, in interview with Kathy Palokoff, May 17, 2017.

4. This quote and following quotes and information from Adam Sobel, in interview with Paul Eder, February 28, 2017.

5. This quote and following quotes and information from Dr. Pernessa Seele, in interview with Kathy Palokoff, February 24, 2017.

6. This quote and following quotes and information from Dr. Kirk Borne, in interview with Paul Eder, April 4, 2017.

7. Susan Lund et al., "Game Changers: Five Opportunities for US Growth and Renewal," McKinsey Global Institute, July 2013, http://www.mckinsey.com/global -themes/americas/us-game-changers (accessed September 18, 2017).

8. Tom Kalil and Fen Zhao, "Unleashing the Power of Big Data," White House President Barack Obama, April 18, 2013, https://obamawhitehouse.archives.gov/blog/2013/04/18/unleashing-power-big-data (accessed September 18, 2017).

9. Thomas Davenport and D. J. Patil, "Data Scientist: The Sexiest Job of the 21st Century," *Harvard Business Review*, October 2012, https://hbr.org/2012/10/data-scientist-the-sexiest-job-of-the-21st-century (accessed September 18, 2017).

10. This quote and following quotes and information from Courtney Scott, in interview with Paul Eder, March 1, 2017.

11. This quote and following quotes and information from Mindy Meads, in interview with Kathy Palokoff, May 6, 2017.

12. This quote and following quotes and information from Louis Lautman, in interview with Cheri Swalwell, February 23, 2017.

13. This quote and following quotes and information from Dr. Barbara Hutchinson, in interview with Kathy Palokoff, February 11, 2017.

14. This quote and following quotes and information from Jerrie Ueberle, in interview with Kathy Palokoff, February 14, 2017.

15. This quote and following quotes and information from David A. Fields, in interview with Kathy Palokoff, March 7, 2017.

16. Malcolm Gladwell, *Outliers: The Story of Success* (New York: Little, Brown, 2008).

17. This quote and following quotes and information from Dame Shellie Hunt, in interview with Raoul Davis, April 19, 2017.

18. This quote and following quotes and information from Rodney Adkins, in interview with Raoul Davis, April 19, 2017.

CHAPTER 28. INSTIGATORS DISRUPT THINGS

1. Patrick Ip, in interview with Kathy Palokoff, January 19, 2017.

2. Barack Obama, "Barack Obama's Feb. 5 Speech" (speech, Barack Obama Campaign Headquarters, Chicago, February 5, 2008), http://www.nytimes.com/2008/02/05/us/politics/05text-obama.html (accessed August 16, 2017).

3. This quote and following quotes and information from Noah Galloway, in interview with Kathy Palokoff, April 12, 2017.

4. This quote and following quotes and information from Marsha Firestone, in interview with Kathy Palokoff, February 21, 2017.

5. Women Presidents' Organization, "WPO Fact Sheet," https://www.womenpresidentsorg.com/about/facts (accessed September 18, 2017).

6. This quote and following quotes and information from David Egan, in interview with Paul Eder, February 22, 2017.

7. This quote and following quotes and information from Yasmine El Baggari, in interview with Kathy Palokoff, May 15, 2017.

8. This quote and following quotes and information from Don Miguel Ruiz Jr., in interview with Kathy Palokoff, January 17, 2017.

9. "Soul to Soul with don Miguel Ruiz," interview by Oprah Winfrey, April 3, 2013, video, 1:44, http://www.oprah.com/own-super-soul-sunday/soul-to-soul-with -don-miguel-ruiz-video (accessed September 18, 2017); Doug Most, "Why Does Tom Brady Read This Book Every Year?" *Boston Globe*, September 8, 2015, https://www .bostonglobe.com/lifestyle/2015/09/08/why-does-tom-brady-read-this-book-every-year/ v7RfxD7FhLNSWezhujIqDI/story.html (accessed September 18, 2017).

10. This quote and following quotes and information from Juliana Richards, in interview with Kathy Palokoff, March 7, 2017 and March 21, 2017.

11. This quote and following quotes and information from Scott Petinga, in interview with Paul Eder, March 1, 2017.

12. This quote and following quotes and information from Dominique McGowan, in interview with Kathy Palokoff, April 16, 2017.

13. This quote and following quotes and information from Ziad K. Abdelnour, in interview with Raoul Davis, February 9, 2017.

14. This quote and following quotes and information from LaToyia Dennis, in interview with Raoul Davis, April 11, 2017.

15. This quote and following quotes and information from Patrick Ip, in interview with Kathy Palokoff, January 19, 2017.

16. This quote and following quotes and information from Karen Benjamin, in interview with Kathy Palokoff, March 7, 2017.

17. Joe Morone, in interview with Kathy Palokoff, March 7, 2017.

18. This quote and following quotes and information from Larry Boyer, in interview with Cheri Swalwell, March 14, 2017.

19. This quote and following quotes and information from Dr. Angela Marshall, in interview with Kathy Palokoff, January 20, 2017.

20. This quote and following quotes and information from Ezz Eldin El Nattar, in interview with Cheri Swalwell, April 26, 2017.

21. This quote and following quotes and information from John Salmons, in interview with Cheri Swalwell, March 7, 2017.

CHAPTER 29. INITIATORS START THINGS

1. Seth Godin, *Poke the Box* (New York: Portfolio, 2015).

2. This quote and following quotes and information from Ellen Kullman, in inter-view with Kathy Palokoff, May 1, 2017.

3. This quote and following quotes and information from Keith Nolan, in written interview with Paul Eder, March 14, 2017.

4. This quote and following quotes and information from Kim Nelson, in interview with Kathy Palokoff, February 23, 2017.

5. Samantha Lile, "7 of the Biggest 'Shark Tank' Success Stories," WallStreetInsanity .com, March 6, 2014, http://wallstreetinsanity.com/7-of-the-biggest-shark-tank-success -stories/ (accessed August 12, 2017).

6. This quote and following quotes and information from David Weild, in interview with Raoul Davis, April 4, 2017.

7. Russ Garland, "Movement Aims to Rally Investors to Fix IPO Market," *Wall Street Journal*, January 13, 2012, https://blogs.wsj.com/venturecapital/2012/01/13/ movement-aims-to-rally-investors-to-fix-ipo-market/ (accessed September 15, 2017).

8. This quote and following quotes and information from Noel Shu, in interview with Kathy Palokoff, March 3, 2017.

9. Merilee Kern, "Maurice Vendôme Champagne Makes American Debut," *FINE Magazine*, January 2016, http://www.finehomesandliving.com/Maurice-Vendme -Champagne-Makes-American-Debut (accessed August 12, 2017).

10. This quote and following quotes and information from Aleen Zakka, in interview with Kathy Palokoff, April 17, 2017.

11. This quote and following quotes and information from Joe Edwardsen, in interview with Kathy Palokoff, January 1, 2017.

12. Soldier, in email to Joe Squared, October 22, 2015.

13. This quote and following quotes and information from Caroline Tsay, in interview with Paul Eder, March 2, 2017.

14. This quote and following quotes and information from Dr. Anton Robert Berzins, in interview with Paul Eder, February 8, 2017.

15. This quote and following quotes and information from Heidi Trost, in interview with Kathy Palokoff, May 28, 2017.

16. This quote and following quotes and information from Rosie O'Gorman and Frank Abruzzese, in interview with Kathy Palokoff, April 20, 2017.

17. "Hezekiah Griggs III," 2013–2014, http://hezekiahgriggs.com/HG3-media kit1314.pdf (accessed September 18, 2017).

18. Hezekiah Griggs III, "Hezekiah Griggs III @ First Baptist Church of Glenarden (2 of 4)," YouTube video, 3:16, presentation to First Baptist Church of Glenarden, posted by "Hezekiah Griggs," October 23, 2014, https://www.youtube.com/ watch?v=PwOT6mu0ExE (accessed August 16, 2017).

19. Ibid.

20. Hezekiah Griggs III, in personal interview with Raoul Davis, January 2010.

21. From Hezekiah Griggs III's "Homegoing Celebration Program," 2016.

22. Griggs III, in personal interview with Raoul Davis.

CHAPTER 30. SUPPORTING AND EXTINGUISHING FIRESTARTERS

1. Joe Mancuso, in personal conversation with Raoul Davis, January 11, 2017.
2. John Sculley, in interview with the author, March 29, 2017.
3. Women Presidents' Organization, "WPO Fact Sheet," https://www.women presidentsorg.com/about/facts (accessed September 18, 2017).
4. Paradigm for Parity, "About the Coalition," https://www.paradigm4parity.com/about#who-we-are (accessed September 18, 2017).
5. Ibid.
6. *Encyclopedia Britannica Online*, s.v. "Hernán Cortés, Marqués del Valle de Oaxaca: Spanish Conquistador," by Ralph Hammond Innes, last modified May 12, 2017, https://www.britannica.com/biography/Hernan-Cortes-marques-del-Valle-de-Oaxaca (accessed June 24, 2017).
7. "What Is the Tony Robbins Firewalk?" Tony Robbins Firewalk, http://www.tonyrobbinsfirewalk.com/what-is-the-tony-robbins-firewalk/ (accessed September 18, 2017).
8. "Salman Khan," Biography.com, last updated April 2, 2014, https://www.biography.com/people/salman-khan-21416751 (accessed June 24, 2017).
9. Dr. Kirk Borne, in interview with Paul Eder, April 4, 2017.
10. *Ironman*, directed by Jon Favreau (Lone Pine, CA: Paramount Pictures/Marvel Studios, 2008).

CHAPTER 31. UNDERSTANDING EMISSIONS

1. "Star Trek Replicator Nearing Reality?" StarTrek.com, January 28, 2014, http://www.startrek.com/article/star-trek-replicator-nearing-reality (accessed September 18, 2017).
2. Larry Boyer, in interview with Cheri Swalwell, March 14, 2017.
3. Robert Cook, "Rosa Parks: The Backstory That Rarely Gets Told," *Newsweek*, December 1, 2015, http://www.newsweek.com/rosa-parks-back-story-rarely-gets-told-399953 (accessed June 24, 2017).

CHAPTER 32. WHY WE NEED MORE FIRESTARTERS

1. *Wonder Woman*, directed by Patty Jenkins (London, UK: Warner Bros., 2017).
2. Dakin Andone, "In Texas, Not All Heroes Wear Capes," CNN, September 3, 2017, http://www.cnn.com/2017/09/03/us/houston-texas-harvey-heroes-trnd/index.html (accessed September 18, 2017).

3. Charles Messenger, *The Chronological Atlas of World War Two* (New York: Macmillan, 1989), p. 242.

INDEX

Page references in *italics* indicate quotes in an epigraph.